BASIC CONCRETE
ENGINEERING
FOR BUILDERS

by Max Schwartz

CD-ROM included inside the back cover:

Concrete Engineering Software, *Limited Version*
by Integrated Structural Software, Inc.

Craftsman Book Company
6058 Corte del Cedro / P.O. Box 6500 / Carlsbad, CA 92018

Acknowledgments

The author thanks the following companies and organizations for furnishing materials and information used in the preparation of this book.

All portions of the *Uniform Building Code* are reproduced from the 1997 edition ©1997, with the permission of the publisher, the International Conference of Building Officials. However, the ICBO assumes no responsibility for the accuracy or completeness of any summaries provided herein.

American Polysteel LLC
5150-F Edith, NE
Albuquerque, NM 87107

Concrete Reinforcing Steel Institute
933 North Plum Grove Rd.
Schaumburg, IL 60173

David White Instruments
11711 River Lane
Germantown, WI 53022

Gurley Precision Instruments
514 Fulton St.
Troy, NY 12180

Integrated Structural Software, Inc.
155 Dorchester Way
San Francisco, CA 94127

International Conference of
 Building Officials
5360 Workman Mill Rd.
Whittier, CA 90601

Smurfit-Stone Container Corporation
8182 Maryland Ave.
St. Louis, MO 63105

Meadow Steel Products
5110 Santa Fe Rd.
Tampa, FL 33619

Nikon, Inc.
19601 Hamilton Ave.
Torrance, CA 90502

Reward Wall Systems, Inc.
4115 South 87th St.
Omaha, NE 68127

Schwing America, Inc.
5900 Centerville Rd.
St. Paul, MN 55127

Somero Enterprises, Inc.
1000 Somero Dr.
P.O. Box 309
Houghton, MI 49931

Symons Corporation
200 East Touhy Ave.
Des Plaines, IL 60018

Williams Form Engineering Corporation
P.O. Box 7389
Grand Rapids, MI 49510

Library of Congress Cataloging-in-Publication Data

Schwartz, Max, 1922-
     Basic concrete engineering for builders / by Max Schwartz.
          p. cm.
     Includes index.
     ISBN 1-57218-091-9
     1. Concrete construction   I. Title.
TH1461.S39   2000
693'.5--dc21
                                         00-047560

© 2000 Craftsman Book Company

Contents

Chapter 1

Concrete Basics

Concrete is the most durable construction material we have. It can be cast in mass in dams and machine foundations, or with intricate detail in spandrel walls in high-rise office buildings. Concrete can be precast in a factory and delivered to a job structurally-ready to carry loads. It's one of our most fire-resistant construction materials. And it can be used in corrosive environments. But for all its diverse uses, concrete is still just a simple mixture of portland cement, sand, aggregate, water, and admixtures.

In this book, we'll tell you just about everything you need to know about concrete — materials, mixing, placing and curing, forming, foundations, slabs, columns, walls, beams, and girders. We'll also tell you about commercial, industrial, and public works construction. And we'll give you checklists and rules-of-thumb to help you easily design concrete structures for each area of work.

We'll make every effort to describe, in simple terms, how to design most of the items you bid on. Here are some of the questions we'll answer on how to plan and build concrete structures:

- What new methods can you use to estimate concrete strength?
- How do you test concrete walls to make sure they're watertight?
- What's the latest information you need to know about concrete joints?
- What's the tolerance on slab thickness for concrete slabs on grade?
- Do long mixing times reduce the strength of concrete?
- How do you place concrete with a conveyor belt?
- What's the recommended way to remove shores and install new ones?
- What is polymer modified concrete?

- How do you use fiber reinforced plastic (FRP) reinforcement in concrete?
- What's the rule-of-thumb approach to formwork?
- How do you maintain safety, quality, and economy in formwork?

Concrete is one of our most common construction materials. It's permanent, resists decay, moisture, fire, and corrosion, and it's easy to install. But it's also the most unforgiving material. If you make a mistake with concrete, it can cost you a great deal to correct it. For example, if you use the wrong blend of ingredients, the concrete can develop holes. If you don't put reinforcement bars in the right place, a wall may spall or crack. If formwork isn't strong enough, a beam or column may be misshapen. And you're sure to have delays and back charges if you set any anchor bolts in the wrong place.

Foundation work requires precision. If one part of a foundation is wrong, the whole building may be affected. With wood or steel framing, you can remove and replace something that isn't right. With concrete, you'll probably have to demolish what's wrong and start over. Once concrete has cured or set up, it's difficult to correct mistakes. This book will help you avoid the more common mistakes and get the job done right the first time.

Structural Design

Long before there were computers, slide rules, or even safety codes, builders were successfully constructing concrete structures. Ancient master builders and architects knew how to build substantial concrete foundations, retaining walls, and pavements. They used trial-and-error and rule-of-thumb methods and adjusted each design for local conditions. If a design worked and withstood the test of time, it became the standard for the area.

As the construction trades became more organized, public agencies, such as building and road departments, worked to standardize design methods. For example, concrete gravity retaining walls, such as the ones shown in Figures 1-1 and 1-2, were developed after centuries of building stone and rubble retaining walls by trial-and-error. Early builders learned that under normal or dry, sandy loam soil conditions, a wall's weight and shape could keep it from overturning or sliding. A wall's weight resists the soil behind it. And the friction between the soil and a wide footing can keep a wall from sliding.

Here are some rules of thumb for retaining walls and their footings:

- Retained earth behind a gravity retaining wall should be 6 inches below the top of the wall.
- The top of a gravity retaining wall should be at least 9 inches wide.

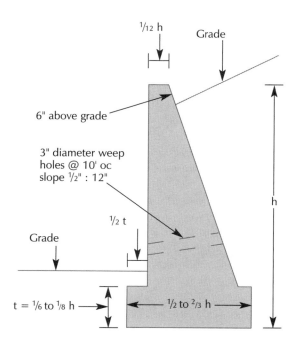

Figure 1-1 Gravity retaining wall with footing

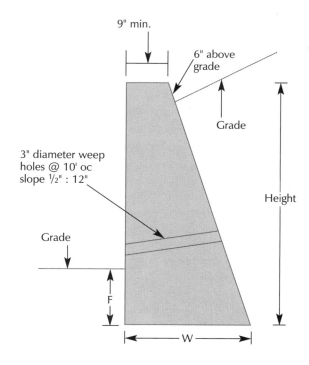

Figure 1-2 Gravity retaining wall without footing

■ A wall should be the same width both below grade and above grade.

■ The width of a gravity retaining wall footing should be $1/2$ to $2/3$ the height of the wall. For example, a footing that's 10 feet below the top of a wall should be 5 to 6.7 feet wide.

■ To avoid surface cracks in a concrete retaining wall, you can place No. 3 bars 2 feet on center both ways and 2 inches from the exposed face of the wall.

■ Install 3- or 4-inch-diameter weep holes 10 feet on center near the lower grade level to reduce hydrostatic pressure on a wall.

■ Use the same rules for cement and stone rubble retaining walls.

■ A footing should be at least $1^1/2$ times as thick as the projection of the footing from the face of the wall. Figure 1-3 shows footing projections and thicknesses for a one- and two-story building.

■ You can use a flared footing instead of an "L" or inverted "T" footing if its effective bearing area is equal to the required rectangular footing. The sides of the flared footing should be at least 60 degrees from the horizontal. A flared footing is commonly used as a pier. Usually, it's cheaper to form a flared footing than a rectangular one. Figure 1-4 shows a simple flared footing.

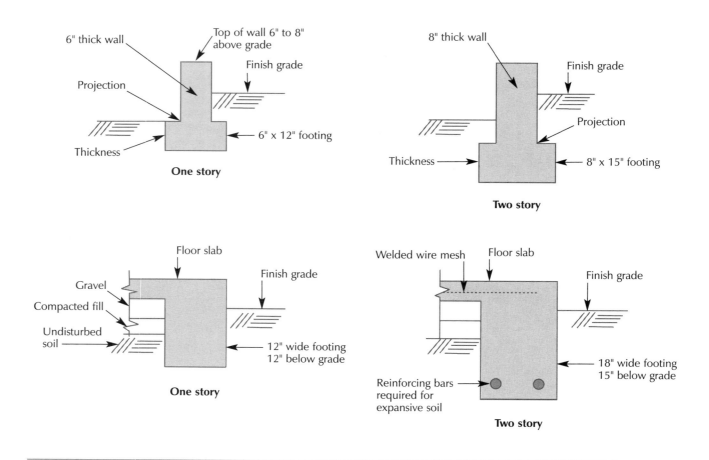

Figure 1-3 Typical dwelling foundations

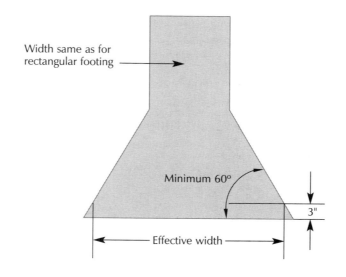

Figure 1-4 Typical flared footing

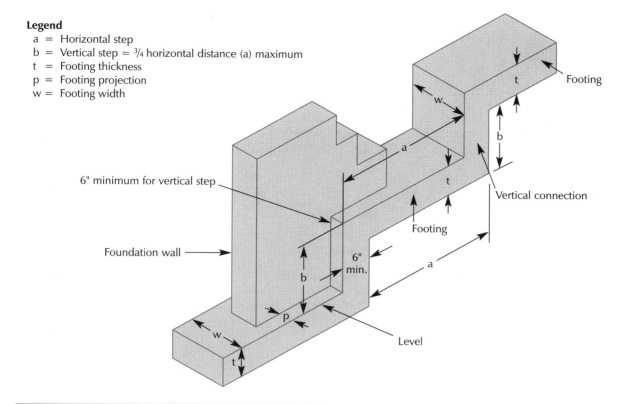

Legend
a = Horizontal step
b = Vertical step = 3/4 horizontal distance (a) maximum
t = Footing thickness
p = Footing projection
w = Footing width

Figure 1-5 Stepped footing

■ When you build a stepped foundation on a sloping grade, height of vertical steps shouldn't exceed ³/₄ of the horizontal distance between the steps. The horizontal distance between the steps should be at least 2 feet. Vertical connections should be the same width as the footing, and at least 6 inches thick. Use the rule shown on Figure 1-5.

Getting Structural Design Information

You can get free structural design from the local building department, concrete industry associations, and public works authorities. For conventional conditions, you can usually rely on their recommended design criteria without having to calculate complex formulas. Here are some sources for free design information:

■ The Federal Housing Administration (FHA) publishes *Minimum Property Standards* which shows minimum designs for conventional construction under normal conditions. This booklet will give you details and specifications for foundations based on conventionally-loaded buildings on average soil conditions (2,000 pounds per square foot or better). The FHA designs are for a dwelling without excessive differential settlement (parts of the building settling at different rates) or movement.

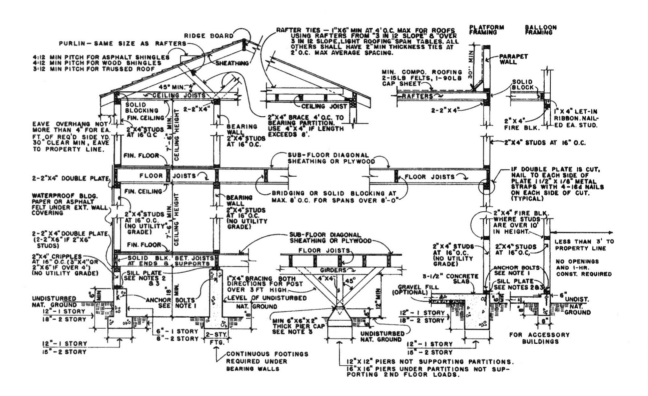

CITY OF LOS ANGELES — DEPARTMENT OF BUILDING AND SAFETY
TYPE V SHEET — WOOD FRAME BUILDINGS
TYPICAL ONE AND TWO STORY CONSTRUCTION DETAILS

B & S B-16
(R. 1-86)

NOTES:

1. Anchor bolts: ½" x 10" bolts embedded 7" in footing wall and spaced @ max. 6' o.c. starting within 12" of sill ends.
2. Footing sills shall have full bearing on the footing wall or mortar bed.
3. All wood in direct contact with masonry or concrete at a point within 48" of the ground shall be pressure treated with an approved preservative or shall be durable wood (Fdn. Gr. Calif. Redwood or Foundation Lumber Western Red Cedar).
4. Under-floor areas shall be ventilated by an approved mechanical means or by openings in exterior foundation walls. Such openings shall have a net area of not less than 1½ square feet for each 25 linear feet of exterior wall. Openings shall be located as close to corners as practicable and shall provide cross ventilation on at least two approximately opposite sides. They shall be covered with corrosion-resistant wire mesh not less than ¼ inch nor more than ½ inch in any dimension.
5. Std. Ftg. Concrete mix: 1 pt. cement, 3 pts. sand, 4 pts. max. 1" rock and max. 8½ gals. water per sack of cement.

Figure 1-6 Standard details for one- and two-story buildings

■ The building trade industry associations are good sources of free design. These include the American Concrete Institute (ACI), Portland Cement Association (PCA), the Concrete Reinforcing Steel Institute (CRSI), and others. These associations publish design manuals that include tables, charts, and drawings of standard structures.

■ The City of Los Angeles Department of Building and Safety hands out, at no charge, *Minimum Standards for Type V Sheet — Wood Frame Buildings and Typical One and Two Story Construction Details*. These standards include exterior and interior foundations, isolated piers, integral foundations, and floor slabs. Most of the foundations built in California for one- and two-story dwellings are based on these documents. Figure 1-6 shows standard details of residential foundations for one- and two-story buildings. The details describe the minimum width, depth, and configuration of various types of footings and foundation walls. The details also call for $1/2$ inch by 10 inch anchor bolts to be embedded 7 inches into the concrete, spaced no more than 6 feet on center, and within 12 inches of the ends of sills.

Any concrete work you do in the public domain is usually regulated by a local Department of Public Works, street department, or similar authority. These agencies will often sell you designs and specifications for concrete sidewalks, curbs, gutters, roadways, retaining walls, and other similar concrete work at a minimal cost. If you use their recommended designs, you usually don't have to submit structural calculations and plans for approval.

Standards for Special Conditions

Because many areas in the United States have special environmental conditions, local building departments adjust standards for their particular soil and weather conditions. For example, builders in Michigan's Upper Peninsula must contend with a 6-foot frost line, so they use a greater footing depth than is needed in other areas. Other building departments, where expansive soil is prevalent, specify wider and deeper footings than conventional footings.

Figure 1-7 shows the relative depth of foundations based on frost line. Figures 1-7A through 1-7F show foundation walls, and Figures 1-7G through 1-7L show area walls. Figures 1-7A and 1-7G are for locations that don't have freezing weather, such as southern California and southern Texas. Figures 1-7B and 1-7H are for areas with minimum temperatures of about 20 degrees F. Figures 1-7C and 1-7I are for areas with minimum temperatures of about 10 degrees F. Figures 1-7D and 1-7J are for New York and the central states, or areas with minimum temperatures of 0 degrees F. Figures 1-7E and 1-7K are for areas with minimum temperatures of minus 10 degrees F, like the North Atlantic states. Figures 1-7F and 1-7L are for areas where the temperature goes below minus 20 degrees F, such as Montana, North Dakota, the northern peninsula of Michigan, and Canada. The

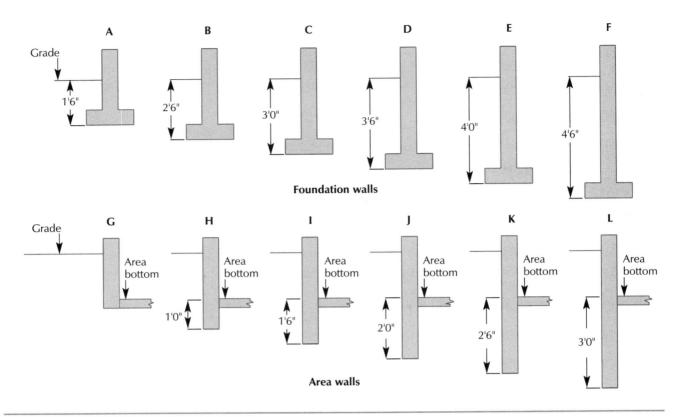

Figure 1-7 *Depth of foundation walls*

minimum temperatures here are minimum temperatures that last for days, not just a minimum temperature that lasts for only a few hours. Figure 1-8 lists the approximate depth of foundations required in some specific locations.

Of course, you should always check with the local building department for the requirements of any project you work on.

Besides city building codes, there are three national and one international basic building codes. They're published by the Southern Building Code Conference (SBCCI), the Building Officials & Code Administrators International (BOCA), and the International Conference of Building Officials (ICBO). The SBCCI code contains special requirements for the southern states, where tornadoes and hurricanes are common. The BOCA code is used in eastern states where there are extreme cold weather conditions. And ICBO's codes, the *International Building Code* (IBC) and the *International Residential Code* (IRC), are published by the International Code Council (ICC).

Special Requirements for Special Industries

Special industries, such as petroleum, mining, and chemical plants, have developed their own standard designs. Here, too, the designs are the result of trial-and-error until universal acceptance made them standards. The major oil

Location	Footing depth below grade	Location	Footing depth below grade
Atlanta, GA	1'6"	Kansas City, MO	3'
Baltimore, MD	3'	Louisville, KY	2'6"
Boston, MA	4'	Milwaukee, WI	5'
Butte, MN	3'	New York, NY	4'
Chicago, IL	4'	Omaha, NE	3'
Denver, CO	1'6"	Philadelphia, PA	3'
Detroit, MI	3'6"	St. Louis, MO	2'
Halifax, Canada	4'	St. Paul, MN	4'
Jackson, FL	1'	Seattle, WA	1'6"

Figure 1-8 *Approximate depth required for foundations*

companies give their facility engineers manuals of standards that show acceptable foundations for pumps, compressors, tanks, and horizontal and vertical vessels. For example, you have to make a foundation for a machine large enough to safely enclose the base of the machine. And you must set anchor bolts deep enough to bear on undisturbed soil. Also, the weight of a machine's foundation should be 3 to 5 times the weight of the machine it supports to resist the machine's vibration.

The mining industry has similar standard manuals for foundations that support large dynamic machines such as rock crushers, rotating mills, and kilns. We'll give you more examples of industry standards later in this book.

Handling Prefabricated Building Items

You don't handle precast structural building items you use on a job the same way as one you build yourself. These include precast concrete planks, double-tee-beams, channel-beams, columns, wall panels, and many types of precast vaults, manholes, and junction boxes. Since a building inspector can't see how an item was built, he or she must rely on preapproval of the product. That's why these items are pre-engineered and inspected. Building codes have very specific rules covering prefabricated structural items. These rules control the design, testing, fabrication, and even periodic inspection of the manufacturer. When you contract for the installation of a prefabricated product, the engineering of that product has already been done. The load capacity and spans are listed in the manufacturer's catalog. The sales representatives should give you any additional assistance you need. A precast item manufacturer will usually use his own cranes and rigging to install an item. We'll go more into this in Chapters 8 and 9.

Another area in which engineering is provided with the product is the lifting hardware for tilt-up wall panels. The subcontractor that lifts and places precast wall panels, beams, and girders will determine the type and location of the lifting hardware. He'll also tell you where to locate inserts in the floor slab to brace the panels. This keeps the responsibility for lifting and placing wall panels without an accident on just one party. The erection subcontractor will select the size of the crane, types of strongbacks, equalizing rigs, and rigging to be used. Strongbacks are steel beams you use to keep a concrete slab from bending while you raise the slab. An equalizing rig is a special arrangement of steel ropes and pulleys you use to make sure you have an equal pull on all lifting cables. Rigging is the system of steel ropes, pulleys, and other lifting devices you use to raise and place precast concrete panels.

Suppliers and Subcontractors

Many suppliers and specialty contractors serve various parts of concrete construction. Depending on the size of a job, the general contractor, specialty contractor, or subcontractor will be responsible for purchasing products or services required for the job. Here are some of the items and tasks involved in concrete construction:

- admixtures for concrete
- batch plants and controls
- bonding adhesives
- breaking, sawing, and cutting concrete
- caissons
- cellular concrete
- coloring concrete
- curbs and gutters
- finishing concrete
- floor coatings
- forms
- foundations
- grinding, scarifying, and grooving concrete
- inspection and testing laboratories
- lightweight concrete
- manholes, vaults, and junction boxes
- mud jacking and slab jacking
- patching and repair

- pavers

- piling, precast and cast-in-place

- pipes

- placing and pumping concrete

- post tensioning

- precast and prestressed planks, beams, and columns

- ready-mixed concrete

- reinforcing steel and fibers

- roadways, driveways, and sidewalks

- septic tanks and clarifiers

- stamping

- vibrating

You can find information on these, and many more specialties, in *The Blue Book of Building and Construction* published by Contractors Register, Inc. This book covers twelve regions of the United States. You can contact them by phone (800-431-2584) or check out their Web site at www.thebluebook.com. Online, you can do a key word or geographic search to find information on a specialty. Or you can use the CSI Master Format headings to get a detailed profile of the advertisers.

Contract Documents

Contract documents include the design drawings, usually prepared by the project architect or engineer, and the placing drawings, commonly drawn by the reinforcing steel fabricator or installer. The placing drawings are based on the design requirements specified on the design drawings, and also describe the amount and size of reinforcing steel that will be supplied and installed.

Design Drawings

On concrete jobs the plans may be called *concrete design drawings*. If the project includes other types of work, the plans are just called design drawings. Concrete design drawings, just like structural steel, electrical and plumbing drawings, have their own special symbols and nomenclature. The form builder, reinforcement bar fabricator, and installer will use the plans, elevations, details, and general notes. In some cases, the job specifications are more important than the design drawings. There's an old saying that "Plans are read at the job site — specifications are read in court."

Here are some rules you should follow when you prepare concrete drawings:

- Show the reinforcing bars in bold lines so they'll stand out. Remember that showing the reinforcement is the main purpose of the drawings.

- Leave out architectural details that may confuse the reader.

- Straight bars, bent bars, ties, and stirrups must fit in the forms in their proper locations.

- Specify clearance for beams and columns to ties or to main steel.

- Show required distances between bars where beams, girders, or joists intersect.

- Dimension the positions of all bars.

- Show all footing elevations.

- Indicate length of caissons.

- Where either straight or bent (or straight only) bars are alternated, indicate the final spacing. For example, #5 @ 12" means to alternate straight and bent bars to provide a 6" spacing between each bar.

- Show where truss bars are bent, i.e. at 1/5, 1/6 point, or dimension the bends. A truss bar is a longitudinal bar that's bent so the ends are near the top of the beam and the center part is near the bottom of the beam. They're placed in the sections of the beam that have the highest tension stress. If a beam is divided into 5 or 6 equal parts, the location of the first point is called the 1/5 point or the 1/6 point.

- Show all concrete that has to be reinforced.

- Don't call for bar shapes that standard fabricating equipment can't make. For special conditions, call the fabricator.

Placement Drawings

Placement or placing drawings are your last chance to check for errors in the design drawings, or in the reinforcing steel subcontract. You should also check to see whether the steel fabricator has changed the sizes or placement of reinforcing steel. It isn't unusual for a fabricator to substitute bars that are in stock for those specified on the design drawings. Changes in bar size usually mean changes in bar spacing to provide equivalent reinforcement.

When a builder, architect, or engineer approves the placement drawings for a job, he or she is responsible for the accuracy and suitability of the reinforced concrete at the job. The fabricator won't begin cutting and bending the reinforcing steel bars in his shop until he receives approved placement drawings.

In summary, the approval of the placement drawings protects the fabricator, the steel bar installer, and the builder. Approved placement drawings also assure the designer that the design drawings were properly interpreted.

This chapter has covered a little of the long history of concrete construction. Its use predates computers, slide rules or building codes. For thousands of years, master builders constructed successful structures using just the rule of thumb. Gradually, master builders matured into general and specialty contractors. And while the principles of concrete mixing, placing and curing have remained more or less the same, machines have replaced manual labor and steel reinforcement has led to thinner and lighter structures.

The rest of this book will describe in detail today's basic rules of concrete design and construction

Chapter 2

Concrete Materials

Cement and concrete aren't the same. There's no such thing as a *cement* sidewalk. Cement is only one of the ingredients that goes into concrete. You need a mixture of cement, aggregate, water, and admixtures, plus labor and time to make a *concrete* sidewalk.

Concrete is a solid blend of cement, sand, gravel (or crushed rock), water, and admixtures. The gravel and sand are called the coarse and fine aggregate. Cement, properly known as portland cement, holds the aggregate together. Figure 2-1 shows how cement, sand, and coarse aggregate should be uniformly dispersed in concrete. The best concrete mixture is a dense one. All the spaces between the crushed rock in a mix should be filled. The cementing and bonding properties of cement are due to the chemical reaction of cement and water. This requires time and favorable conditions, such as temperature and moisture. Let's look more at each ingredient in concrete.

Cement

Joseph Aspdin, an English stone mason, patented portland cement in 1824. He named it after the gray stone found on the Isle of Portland. Cement is made from limestone, which is mainly a calcium carbonate. Limestone is mined and processed by crushing and roasting it with clay in rotating kilns. A typical cement kiln has a 15-foot-diameter shell that slopes slightly from the feed end to the discharge end, and rotates slowly in a clockwise direction. The kiln normally operates at 3,000 degrees F. This produces a dehydrated porous material called *clinkers*. The clinkers are then ground into a fine powder to become portland cement.

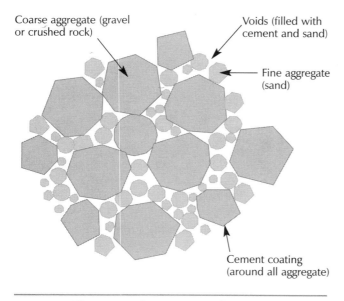

Figure 2-1 *Components of concrete*

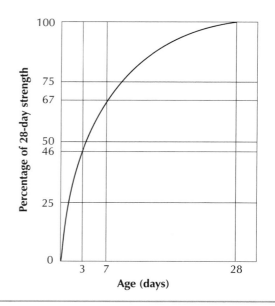

Figure 2-2 *Normal concrete strength vs. time*

Portland cement is usually shipped in cloth or paper bags that contain 1 cubic foot of loose volume and weigh about 94 pounds each. A cubic yard of concrete contains 27 cubic feet and weighs about 2 tons. Large quantities of cement are sometimes referred to as barrels. A barrel is equivalent to four 1-cubic-foot sacks, and weighs about 376 pounds. Portland cement is also shipped in bulk by rail or trucks with specially-designed hopper cars.

Types of Cement

There are five common types of portland cement — Type I normal, Type II modified, Type III high early, Type IV low heat, and Type V sulfate resistant. All portland cement should conform to American Society of Testing Materials (ASTM) C-150 or C-595 standards, and building codes refer to these standards. The most common standards are:

- ■ *Cement:* ASTM C-150 and C-595
- ■ *Aggregates:* ASTM C-33 and C-330
- ■ *Water:* ASTM C-94
- ■ *Admixtures:* ASTM C-10, C-260, C494, C-595, and C-618
- ■ *Batching:* ASTM C94 and C-685

Type I normal portland cement is used for general construction where you don't need the special properties of the other types. Use it in pavements and buildings that aren't exposed to acids or salts. This type of concrete reaches nearly its full strength in 28 days. Figure 2-2 shows the relationship between curing time and strength of Type I normal cement.

If a structure you're building will be exposed to acids or salts, use Type II modified cement. If you're working on a beachfront building that must resist salty mist, this is the cement you'd use. Type II modified cement cures more slowly than Type I normal. In 28 days it's about 85 percent the strength of Type I normal cement. It may take 3 months to reach its full strength.

If you need concrete that cures quickly, use Type III high early cement. You'll need this when traffic or access to a building must wait until the concrete is cured. This type of cement is 190 percent stronger than Type I normal cement in 3 days, 130 percent stronger in 28 days, and 115 percent stronger in 3 months.

Most building contractors don't need Type IV low heat cement. It's used mainly for massive pours, such as a dam, where a mixture will generate a lot of heat as it cures.

Use Type V sulfate resistant cement if a structure will be directly exposed to salt water or other severe conditions. This cement cures slowly, reaching only 65 percent the strength of Type I normal cement in 28 days, and 85 percent in 3 months. Figure 2-3 lists the relative strength of different types of cement as a percentage of the strength Type I normal cement would have in a particular number of days. For example, concrete made with Type I normal cement may have a compressive strength of 2,000 psi in 28 days, 1,340 psi (67 percent of 2,000) in 7 days, and 920 psi (46 percent of 2,000) in 3 days. A mixture made with the same amount of Type II modified cement would have a compressive strength of 1,700 psi (85 percent of 2,000) in 28 days, and 736 psi (80 percent of 920) in 3 days. If the blend is made with Type II modified, it'll take 3 months before it reaches a strength of 2,000 psi.

On the other hand, the same concrete mix made with Type III high early cement has 190 percent more strength than Type I normal cement has in 3 days, or 1,748 psi (190 percent of 920). It'll be 130 percent stronger than Type I normal cement in 28 days (130 percent of 2,000), or 2,600 psi.

Type	3 days	7 days	28 days	3 months
I	100%	100%	100%	100%
II	80%	82%	85%	100%
III	190%	120%	130%	115%
IV	50%	55%	65%	90%
V	65%	75%	65%	85%

Figure 2-3 Approximate relative strength of five types of cement

Other types of portland cement are air-entrained portland cement, portland blast furnace cement, natural cement, white portland cement, waterproof portland cement, and oil well portland cement. Specifications for air-entrained portland cement are given in ASTM C-175. Portland blast furnace cement is described in ASTM C-205, and natural cement is specified in ASTM C-10. White portland cement is made from selected materials. To make waterproof portland cement, water-repellent materials are added to the clinkers before they're ground into powder. Oil well portland cement is made to harden properly at high temperatures. It's used for very deep oil wells.

Ratio of Cement to Aggregates

The ratio of portland cement to sand and rock in a mix directly affects its strength and cost. For field-mixed concrete, ratios of sacks of cement per cubic yard of concrete are 6.0, 5.8, 5.4, and 5.2. Adding more cement produces higher-strength concrete. The relationship of rock size, sacks of cement, volume of water, sand, and crushed rock is given in Figure 2-4. These values are for structures that aren't exposed to freezing or thawing.

Normally, the ratio of a mix is given in sacks of cement per cubic yard of concrete. The ratio of cement to concrete is also given by weight. Use a 5-sack mix to get 2,500 to 3,000 psi concrete for foundations, walls, and footings. Use a 6-sack mix to get 3,500 to 4,000 psi concrete for driveways, slabs, and columns. For highly-detailed architectural items, such as concrete ledges, sills, and posts where you need 4,500 and 5,000 psi concrete, use a 7-sack mix.

A lean, stiff mixture with large crushed rock is 15 percent water, 7 percent cement, and 78 percent crushed rock. A wet mixture with small gravel is 20 percent water, 14 percent cement, and 66 percent aggregate.

Crushed rock is made from different types of stone. These include limonite, barite, and magnetite. Magnetite is the densest crushed rock. In special cases, coarse aggregate is made of scrap iron, punchings, and cast iron. Punchings are the waste produced when steel plates or sheets are perforated.

Maximum rock size (in)	Cement (sacks per cu yd)	Water per sack of cement (gal)	Sand per sack of cement (pounds)	Rock per sack of cement (pounds)
3/4	6.0	5	170	230
1	5.8	5	160	255
1 1/2	5.4	5	150	300
2	5.2	5	150	335

Figure 2-4 Ratio of cement to water, sand and rock in field-mixed concrete

Water

The key to properly-mixed concrete is water. Water changes a loose mixture into a rocklike solid material. When you mix water with cement, it makes a paste. Some of the water joins chemically with the cement and starts a chemical reaction. The paste will set rapidly at first, then more slowly for a long time. When the chemical conversion is complete, or cured, the composite will be rock-hard. Cement paste is between 22 and 34 percent of the total volume of concrete.

To mix concrete, use clean drinkable (potable) water, if you can. If in doubt, have the water you use analyzed. Make sure it's free of acids, alkalis, oil, and decayed vegetable matter. All of these may interfere with your concrete setting up.

You can use seawater to mix concrete, but its strength will be 10 to 20 percent lower than if you use fresh water. You can offset this by using less water in the mix or increasing the amount of cement. Seawater won't deteriorate steel reinforcement if the steel isn't exposed to air. But the combination of salt water and air will corrode steel rapidly and cause concrete to spall.

The amount of water in a mix also determines its compressive strength. Measure the water you add to a sack of cement in gallons. Use 4 gallons of water per sack of cement for high-strength concrete. Use $8\frac{1}{2}$ gallons of water per sack of cement for a low-strength mix. Some authorities recommend that you don't add more than $7\frac{1}{2}$ gallons of water per sack of cement, including free water. Free water is the water that's already in any aggregates and admixtures in the mix. Figure 2-5 shows the relationship between compressive strength and water/cement ratios by weight.

You need only a little water to start a chemical reaction in cement, but you have to add more water to be able to work with it. If you don't use enough water, a mixture will be too stiff to work with, and too much water will make it weak. Also the cement may wash out of the mixture. When too much water evaporates out of a mixture, you'll get air pockets. Try to use as little water as possible and still have a workable mixture.

Compressive strength (psi)	Water/cement ratio by weight
2000	0.80
3000	0.69
4000	0.57
5000	0.47
6000	0.40

Figure 2-5 *Compressive strength vs. water/cement ratio*

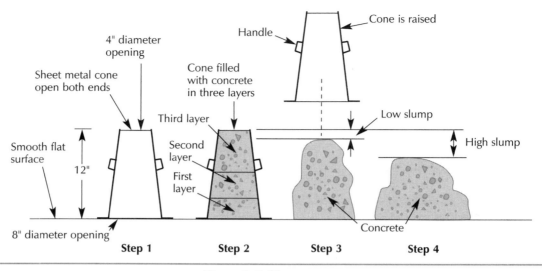

Figure 2-6 Slump test

Slump Tests

Use a slump test to measure how easy a mixture will be to work with. It will tell you how fluid, and how soft or wet, a batch of concrete is. Figure 2-6 shows how to make a slump test using a 12-inch-high sheet metal cone that's open at both ends. The bottom of the cone is 8 inches in diameter, and the top is 4 inches in diameter.

Start the slump test by moistening the inside of the metal cone. Then place the cone on a flat smooth surface, with the large end down, and fill it with fresh concrete in three layers. Tamp each layer before you add the next. After the cone is completely full, lift the cone so the mixture slumps down. Measure the height of the mixture. Subtract this height from the 12-inch height of the cone to give the result of the slump test. If the number is high, the mixture is a wet mix. If it's low, the concrete is a dry mix. Recommended slumps for different types of work are shown in Figure 2-7. If the fresh concrete isn't thoroughly vibrated (tamped), increase the values in the table by 50 percent. A slump should never be more than 6 inches. Generally, a mix can be described as stiff, medium, or wet. Use a stiff mix

Type of concrete work	Maximum slump (inches)	Minimum slump (inches)
Reinforced foundations	4	2
Plain foundations and caissons	3	1
Concrete slabs, beams, walls, and columns	5	2
Pavements	2	1
Mass concrete	2	1

Figure 2-7 Maximum and minimum concrete slumps

for foundation walls, pavements and floor slabs. Use a medium mix for tank walls, some floor slabs, and beams. Use a wet mix for very thin, heavily reinforced walls. A stiff mix has a slump number between 0 and 3. A medium mix is between 3 and 6, and a wet mix is between 6 and 8.

Aggregates

All aggregate should be clean hard rock with no attached dirt, silt, clay, coal, mica, salts, humus, or other organic material. Also, don't use friable or laminated aggregate such as shale or chert. The surface of the aggregate particles should be angular and sharp. Don't use round smooth aggregate particles, such as beach sand, because they don't stick together well.

There are two general weights of aggregate — regular and lightweight. Regular-weight aggregate is made from crushed granite rock. Lightweight aggregate has air pockets. It's made from volcanic material, or by roasting blast furnace slag, shale, perlite, or slate.

Aggregate, gravel, or crushed stone larger than $1/4$ inch across is considered coarse. Fine aggregate, which includes sand, is smaller than $1/4$ inch across. Coarse aggregate may be classified as No. 1, 2, 3, or 4 gravel, depending on the size of the particles. Sand may be natural, or manufactured by crushing rock. Sands are classified as coarse, medium, or fine. Natural aggregate should conform to ASTM C-33 and lightweight aggregate should conform to ASTM C-330.

There's another rule you should follow about the maximum size of coarse aggregate you use. The size of crushed rock shouldn't be more than $3/4$ of the clear spacing between reinforcing bars, or the distance between the bars and forms. In slabs or other flatwork, the aggregate should be no larger than $1/5$ the distance between the forms, or the thickness of the slab, whichever is narrower.

Admixtures

You use admixtures to improve the workability, strength, durability, and water resistance of concrete. Additives can help keep aggregate from separating, reduce the heat of hydration, adjust entrained air, and slow or speed how quickly a mix sets or hardens. You also can use an admixture to color concrete. Some types of admixtures are:

- water-reducing admixtures
- retarding admixtures

- accelerating admixtures
- workability admixtures
- dampproofing admixtures
- permeability agents
- gas forming agents
- coloring agents

Powdered materials such as diatomaceous earth, pumice, fly ash, hydrated lime, and other fine aggregates increase the workability of a concrete mix. You can use carbohydrates to reduce the amount of water you need in a mix. Lignin, borax, sugars, and tartaric acids retard curing, while calcium chloride speeds up curing. You can use calcium soaps to dampproof concrete. To increase the amount of entrained air in a mix, use wood resin salts or ammonia stearates that are gas formers. But be sure any additives you use are approved, *in writing*, by the engineer on the job, or by the owner.

To harden the surface of a slab, sprinkle an admixture (such as Masterplate) or finely-ground iron particles on it. Masterplate is a common surface hardener, manufactured by Master Builders, that contains iron powder. You trowel the iron particles into the surface to make the slab more resistant to abrasion and wear. Do this just before you float the surface, and again before you do the final trowel work. You can also use Masterplate in warehouses where there's lots of traffic with steel wheel equipment.

Be careful with admixtures that have lots of chloride or salts. Don't use more than 2 percent calcium chloride in a mix, as measured by the weight of the cement in the mix. It can corrode steel reinforcing bars, which in turn can cause spalling. Always check on the chloride content of an admixture.

To get more strength by reducing the amount of water and still maintain workability, use a pozzolan admixture. This material increases the slump of a mixture up to 150 percent without adding more water. Make sure any pozzolan admixture you use conforms to ASTM C-618. The benefits of a pozzolan admixture are that it:

- makes concrete easier to place
- increases the workability of concrete without increasing the water-cement ratio, which reduces the strength. In fact, a pozzolan admixture may increase the strength of a mix.
- makes concrete easier to finish
- gives concrete a better finished appearance

You can use admixtures to improve the entrained air in a mix. Air-entrained concrete mix isn't the same as air trapped in the mixture. Trapped air is air in relatively large air voids that aren't dispersed uniformly throughout a mix. Entrained air is made up of minute disconnected bubbles that are well distributed throughout a mix.

Too little entrained-air admixture in concrete won't protect concrete against cyclic freezing and thawing. It can lower the strength of a batch of concrete by 9 percent in severe exposures and 7 percent in moderate exposures. Using too much entrained-air admixture will also reduce the strength of concrete. The amount of air-entraining admixture you should use depends on the type of cement, mix proportions, slump, aggregate, type of mixer, mixing time, temperature, and other factors. You'll usually have to make some trial-and-error batches. Generally, the proportion of sand to total aggregate can be about $4^1/_2$ percent of entrained-air admixture for a mix that uses $1^1/_2$- to $2^1/_2$-inch aggregate. Concrete is easier to work with when there's a larger proportion of mortar in the mix, which increases its cohesiveness. Batch plants normally consider these factors when designing a mix.

You should test the air content of a batch of concrete frequently. There are several methods you can use to do this — the volumetric method (ASTM C-173), the pressure method (ASTM C-231), or the gravimetric or weight method (ASTM C-138). For lightweight concrete, use the volumetric method. We'll give you more information on these tests in the next chapter.

Use a pure mineral pigment admixture to color concrete. Any coloring material should remain chemically stable in alkaline conditions, such as exposure to seawater. The coloring pigment shouldn't fade in sunlight, and it shouldn't affect the strength and curing of the concrete. It's a good idea to make several trial runs with a new color admixture. Usually you use less than 10 percent pigment in concrete. About 3 to 6 pounds of high grade pigment per sack of cement is usually enough. For light colors, you can use some or all white cement to obtain the shade and clarity you want. Here are some types of admixtures you can use to get certain colors:

- *grays and blacks:* black iron oxide, mineral black, and carbon black

- *browns and reds:* red iron oxide, brown iron oxide, raw and burnt umber

- *blues:* cobalt blue, ultramarine blue, phthalocyanine blue

- *greens:* chromium oxide, phthalocyanine green

- *ivory, cream, or buff:* yellow iron oxide

White cement isn't an admixture. It's a special type of portland cement with the same properties as regular cement, except it's white. You should use white or light-colored aggregate with white portland cement. Don't use form oil, because it may the discolor the concrete.

Mortar and Grout

Many in the construction business confuse mortar and grout. And a mason and a tile setter don't use the same terms. In this book, mortar is a mix of cement, lime, water, and fine aggregate, but without coarse aggregate. Grout is a mixture of cement and water, either with or without fine aggregate. Grout is a wetter mixture than mortar. You use grout to fill cells in hollow masonry or cavity brick walls. You use mortar to fill joints between masonry units. Specifications for grout are covered in the *Uniform Building Code* Standard 21-18. Specifications for mortar are in *UBC* Standard 21-15.

There are four types of mortar: M, S, N, and O. Figure 2-8 shows the proportions of their ingredients and their relative compressive strengths.

Lightweight Concrete

Lightweight concrete has many benefits over regular heavyweight concrete. In tall structures, it reduces the weight of a building, saving money on foundations and columns. Also, the structure has less mass that may be affected by earthquake forces. Lightweight concrete provides fireproofing to a building's structural frame. You can use lightweight concrete for roof or floor fill. Vermiculite and perlite concrete make good insulators, but you shouldn't use them for structural purposes. You can use lightweight structural concrete for:

- reinforced concrete frames of multistory buildings
- bridge decks
- prestressed and poststressed concrete structural elements
- concrete pipes
- concrete masonry units
- Gunite™ applications
- precast concrete elements

Type	Portland cement	Masonry cement	Lime	Strength (psi)
M	1	1	$^1/_4$	2,000
S	1	$^1/_2$	$^1/_4$ - $^1/_2$	1,800
N	1	-	$^1/_2$ - $1^1/_4$	750
O	1	-	$1^1/_4$ - $2^1/_2$	350
Note: Don't use less than $2^1/_2$ times, or more than 3 times, aggregate than cement plus lime.				

Figure 2-8 Ingredients and relative strength of mortar types

Lightweight concrete made of expanded shale has the following advantages over ordinary concrete:

- It's one-third lighter.
- It absorbs less moisture.
- It shrinks less as it cures.
- It has higher compressive strength.
- It's a fire retardant.
- It has a higher insulation value.

Lightweight Concrete Aggregates

Aggregate for lightweight concrete may be made from vermiculite (a type of mica), perlite (a volcanic rock), or expanded clay. Concrete with these aggregates is sometimes called lightweight insulating concrete. Vermiculite ore, when exposed to high temperatures, will expand to 20 times its original size. Use it to make nonstructural concrete that weighs between 20 and 40 pounds per cubic foot. This is much lighter than regular concrete, which weighs about 150 pounds per cubic foot.

Perlite also expands when treated with water and heat. Lightweight concrete made with perlite weighs about 85 pounds per cubic foot.

After clay is crushed and screened, it's roasted in a kiln, retort, or furnace to 2,000 degrees F, where its particle size expands to about $3/4$ inch. This produces a strong and durable material that's chemically inert, low in water absorption, and light in weight. It generally weighs between 90 and 100 pounds per cubic foot.

Blast furnace slag can also be used as a coarse aggregate in lightweight concrete. Other volcanic rocks are scoria, pumice, and cinders.

Lightweight insulating concrete may weigh (oven-dry) from 15 to 90 pounds per cubic foot, and its 28-day compressive strength may be from 100 to 1,000 psi. This type of concrete may be grouped as follows:

- *Group I:* made with aggregates of expanded materials like perlite or vermiculite.
- *Group II:* made with aggregates manufactured by expanding, calcining, or sintering blast furnace slag, clay, diatomite, fly ash, shale ore slate, or by processing natural materials like pumice, scoria, or tuff.
- *Group III:* made by adding a preformed or formed-in-place foam to a cement paste or cement-sand mortar. The foam forms millions of air bubbles in the mix.

Cellular Lightweight Concrete

Cellular lightweight concrete is another type of lightweight material. It's made of aerated or foamed concrete or mortar that contains bubbles of air or gas. The most popular use of cellular lightweight concrete is as a floor underlayment in wood-frame multistory buildings. It's an excellent base for a carpet and pad or finish flooring. A $1^1/_2$-inch-thick layer of cellular concrete underlayment installed with a $^5/_8$-inch-thick gypsum wallboard ceiling provides one-hour fire protective rating to wood floor framing. It also provides soundproofing between floors. Cured cellular lightweight concrete has a compressive strength of 1,000 to 1,200 psi.

Reinforcement

Although concrete is strong in compression, it's weak in tension. Without steel reinforcement, it can crack and break. To develop its tensile strength, you need to reinforce it. Reinforced concrete depends on three major phases of work:

- *Concrete mix:* cement, sand, gravel, water, and admixtures
- *Reinforcement:* usually welded wire mesh (also called electric welded wire mesh [EWWM], or wire fabric) or deformed steel bars
- *Formwork and shoring:* sheathing, framing, bracing, and supports

The most common types of steel reinforcement are welded wire fabric and deformed and plain steel bars. Other types of concrete reinforcement are fibers of steel, glass, and plastic. With steel fibers, you add the fibers with a special admixture as you mix the concrete. This should be done in accordance with the steel fiber manufacturer's specifications. When mixed, it will produce concrete that's *elasticplastic*, as defined by ASTM C018.

Welded wire fabric, also called wire mesh, is identified by the gauge and spacing of its wires. For example, $4 \times 12^3/_6$ stands for longitudinal wires 4 inches on center, transverse wires 12 inches on center, No. 3 gauge longitudinal wires, and No. 6 gauge transverse wires. Welded wire fabric comes in lengths of 150 to 200 feet. The width is usually 84 inches. The gauge of wire shows its diameter. We'll give you more information on how to use wire mesh in the chapter on floor and roof slabs. Welded wire reinforcement should conform to ASTM A-185.

Reinforcing bars come in several grades. They were called structural, intermediate, and hard grade. Now they're called:

- Grade 60, standard grade, which has a yield strength of 60,000 psi
- Grade 40, lower strength, which has a yield strength of 40,000 psi
- Grade 75, premium grade (usually not stocked by suppliers) which has a yield strength of 75,000 psi

You can identify a plain undeformed steel bar by its diameter, such as a $1/2$-, $5/8$-, or $3/4$-inch bar. A deformed bar has irregular surfaces to keep it from slipping in concrete. Identify it by a number that stands for its average diameter in eighths of an inch. For example, a No. 5 deformed bar has the same cross-sectional area as a plain $5/8$-inch diameter bar. Figure 2-9 lists the properties of most standard deformed reinforcing bars. Standard mill lengths of reinforcing bars are 20, 40, and 60 feet. You can get nonstandard lengths by special arrangement with a supplier. All reinforcing bars should conform to ASTM A-15 and A-305.

In 1997, several manufacturers began making soft metric reinforcing bars to use in non-metric jobs. Some mills intend to shift completely to producing metric bars, and will reduce or stop making the inch-pound bars. This will depend on their inventory of the old bar, the level of acceptance of the metric bar, and how completely the construction industry adopts the metric system. Federally-funded construction jobs must be designed in metric units and built with metric materials. Private sector jobs don't have to be metric yet. By producing soft metric bars, the manufacturers can satisfy both requirements, but maintaining dual inventories to serve two markets is more difficult and expensive.

We need to become accustomed to the terms of soft metric, hard metric, and inch-pound units as we move into using the metric system. SI Units are units of the International System of Units (SI) and other units specifically approved in ASTM E380. Inch-pound units are based on the inch and pound commonly used in the United States (sometimes called the English System). Soft metric conversion of reinforcing bars describes the dimensions of inch-pound bars in terms of SI metric units, but doesn't physically change the size of the bars. Hard metric conversion sets the sizes of reinforcing bars entirely in SI sizes. This involves re-engineering the sizes and physically changing the bars. Figure 2-10 shows a comparison of soft metric bar sizes to inch-pound bar sizes.

Bar size, designation number	Nominal diameter (inches)	Cross-sectional area (sq inches)	Weight (lbs/ft)
2	$1/4$	0.05	0.167
3	$3/8$	0.11	0.376
4	$1/2$	0.20	0.668
5	$5/8$	0.31	1.043
6	$3/4$	0.44	1.502
7	$7/8$	0.60	2.044
8	1	0.79	2.670
9	$1 1/8$	1.00	3.400
10	$1 1/4$	1.27	4.303
11	$1 3/8$	1.56	5.313

Figure 2-9 Standard reinforcing bars

Metric	Inch-pound
#10	#3
#13	#4
#16	#5
#19	#6
#22	#7
#25	#8
#29	#9
#32	#10
#36	#11
#43	#14
#57	#18

Figure 2-10 Comparison of metric vs. inch-pound bar sizes

Have anyone who designs a reinforced concrete building for you help make the transition to metric by including a conversion table on project drawings or specifications, like the one shown in Figure 2-10. Also designate or label reinforcing bars on drawings and specifications with metric bar sizes. Spacing of bars, lap lengths, and other dimensions can still be given in feet and inches. For example, change #5 @ 12" to #16 @ 12".

Epoxy-Coated Reinforcing Bars

Structures exposed to the weather often deteriorate because their reinforcing bars corrode. Salty water (from the deicing salts) can penetrate the concrete pavement and corrode the steel reinforcing bars. One way to prevent bars from rusting is to use epoxy-coated reinforcing bars. In certain cases, steel reinforcing bars are coated with zinc. Typical structures that are built with coated bars include bridges, wharves, stadiums, open parking buildings, and wastewater treatment plants. Coated bars should conform to ASTM A-15 and A-305.

Coated bars may be fabricated, bent, and installed at a job. But be careful handling coated bars. You don't want to scrape or chip the coating. Use nylon or similar fabric straps to bundle the bars. Normally, you're allowed 3 percent of scraped or chipped area. You can repair damaged areas on epoxy-coated bars in the field with two-component epoxy compound.

Supporting Reinforcing Bars

When you place reinforcement bars in a slab, you need some type of support to hold the bars in place. On very inexpensive slabs on grade, you can place small concrete blocks on the subgrade to support the bars. But in most cases, you'll use chairs and bolsters in both elevated slabs and slabs on grade. Chairs and bolsters are metal accessories made of #7 wire to support reinforcing bars. Most are coated with a plastic on the lower portion of the unit to prevent the unit from rusting.

Metal bar supports are manufactured in three classes, each designed for a particular use and exposure to corrosion. *Class 1* supports are for maximum protection, and where the concrete surface may require light grinding or sandblasting. The support legs are protected with a plastic coating. The steel wires are coated with plastic. *Class 2* supports are for moderate protection and where the concrete surface may require light grinding or sandblasting. The bottom of each leg is protected by a stainless steel tip. *Class 3* supports have no protection over the bright basic wires against rusting. These supports are used where surface blemishes are acceptable or where supports will not be in contact with an exposed concrete surface.

Be sure you use the right kind and height of chair or bolster. There are low and high chairs, especially high chairs, and continuous high chairs. Bolsters are made for lower and higher bars. They're usually stocked in heights from $3/4$ inch to 5 inches. Select support sizes so that the reinforcing bars have the following minimum concrete cover:

- 3 inches for concrete cast against the ground without forms.

- 2 inches for No. 6 bars or larger when concrete is exposed to the weather or ground, but cast against forms.

- $1^1/_2$ inches for No. 5 bars or smaller when concrete is exposed to the weather or ground but cast against forms.

- $1^1/_2$ inches for beams and columns without exposure.

- $3/_4$ inch for slabs and walls without exposure.

Be sure to check your local building code for any special requirements. Then make sure you've installed adequate support for reinforcing bars before you place any concrete around them. Make sure the bars can't move beyond the specified tolerances. You don't have to tie bars at every intersection. In slabs, you should tie bars, each way, every fourth or fifth intersection. In walls, you should tie the bars, each way, every third intersection. Depending on job conditions, you should place bars on wire bar supports, precast concrete bar supports, cementitious fiber reinforced bar supports, all-plastic bar supports, or special chairs for epoxy-coated bars.

Recommended References for Concrete Reinforcing Steel

Here are some useful handbooks on concrete reinforcing steel that are published by the Concrete Reinforcing Steel Institute (CRSI):

Manual of Standard Practice, 1997 edition

Reinforcement Anchorage and Splices, 1997 edition

Reinforcing Bar Detailing, 3rd edition

Field Handling Techniques for Epoxy-Coated Rebar

Placing Reinforcing Bars

Construction of Continuously-Reinforced Concrete Pavements

Embedded Items

Your concrete work must merge smoothly with the other structural and architectural parts of a building. This means you have to set all embedded items in exactly the right place. You fasten most of these items to the inside of your forms before you place any concrete. Here are some of the most common embedded items you should know about:

- *Anchor bolts:* Bent steel rods with a threaded exposed portion. They hold steel columns to a foundation.

- *Flashing reglets:* Horizontal inserts above the roof line. Metal flashing is inserted into these reglets.

- *Stone and brick anchors:* Used to fasten a stone or brick veneer to a concrete wall.

- *Wedge inserts:* Made of malleable iron. Use them to anchor steel framing to concrete walls or columns.

- *Coiled inserts:* Threaded devices that are used in tilt-up construction. Some are set into a floor slab to hold the lower end of a brace. Others are set into a concrete wall to hold the upper end of a brace or to use as lifting points.

Most embedded items are factory fabricated. You just buy them from your concrete hardware supplier. Most have been tested in a laboratory and their load capacity has been proven. But in some cases, the architect or structural engineer will require a special device. Before you have one of these fabricated, be sure to check with the building department to see if they approve of your design.

Sealants and Grouts

To finish a concrete job, you'll probably need sealants and grouts. Use a sealant to protect the expansion and construction joints in slabs on grade from moisture and sunlight. (Construction joints are those made at the edge of one pour and the beginning of another pour.) Common joint sealants are made of polyurethane, polysulfide, or silicone. In expansion and construction joints, sealants remain elastic and protect steel dowels from rusting. You should first install backer rods made of foam material, stuffing them into the upper part of a joint before you apply sealant. You'll find more information on backer rods in the chapter on floor and roof slabs. For concrete basement walls, you'll need a waterproof sealant. To fill cavities in concrete, use a nonmetallic grout that won't shrink.

Be sure to patch defects and tie ends for a good finished job. Unless you seal steel tie ends, they'll corrode and stain the surface of the concrete.

Using the Metric System

Since construction is moving into the metric age, here's a review of how to show metric unit measurements on construction documents:

- Length = meter (m)
- Mass = kilogram (kg)
- Time = second (s)

Figure 2-11 shows how to indicate orders of magnitude. But there can be some confusion because the symbol *m* can mean either meter or milli. It's usually pretty easy to figure out which it is by the context in which it's used. A meter is a unit of length. Milli is a prefix used with SI units, representing the factor 10^{-3} that the units are multiplied by.

As metric terminology becomes more common in construction work, you'll want to follow some rules when you use metric units. Present day reference books and codes don't always follow these rules yet, but probably will in the future. Here are some of the rules for writing metric symbols and quantities:

- Use lower case for unit names.
- Use lower case for unit symbols of 1,000 and less, such as k.
- Use upper case for unit symbols of 1,000,000 and more, such as M and G.
- Don't use compound prefixes, such as nm.
- It's kg, not k g with a space between the k and g.
- Don't mix names and symbols.
- Don't use a period after a symbol, such as 12g., except at the end of a sentence.
- Don't use fractions, use decimals such as 0.75 kg, not $^3/_4$ kg.
- Use a zero before the decimal marker for values less than 1.
- Use a dot to indicate multiplication.
- Use meters and millimeters for construction dimensions.

Prefix	Symbol	Order of magnitude	Number
kilo	k	10^3	1,000
milli	m	10^{-3}	0.001
mega	M	10^6	1,000,000
gigi	G	10^9	1,000,000,000
micro	μ	10^{-6}	0.000001
nano	n	10^{-9}	0.000000001

Figure 2-11 Metric orders of magnitude

- Don't use centimeters.

- Use meters and kilometers for surveying information.

- Use whole numbers to indicate millimeters, 25 mm, and decimal numbers to indicate meters, 3.750 m.

- Use square meters for small areas and square kilometers for large areas.

- Use square millimeters for very small areas.

- Don't use square centimeters.

- For volumes of construction materials, use cubic meters; for fluids, use liters.

For engineering terminology, use the following:

- Mass and force are separate units. The basic unit of mass is the kg. It's independent of gravity. The basic unit of force is the Newton (N). It's equal to mass times acceleration.

- Stress is given as $\dfrac{N}{mm^2}$

- Don't use "weigh" or "weight," use "mass."

Measuring Concrete Materials

Figure 2-12 shows the fundamental units of measurements for concrete mixes.

Adding 1 gallon of water per cubic yard of concrete increases slump about 1 inch. Increasing slump by 1 inch decreases concrete strength about 200 psi.

The specific gravity of a material is the ratio of the weight of the material to the weight of an equal amount of water. You need to know the specific gravity of a material when you want to figure its weight, density, and volume.

Unit weight	Weight of 1 cubic foot of material
Unit weight of cement	94 lbs.
Unit weight of water	62.4 lbs.
1 gallon of water	8.33 lbs.
1 cubic foot of water	$\dfrac{62.4}{8.33} = 7.5$ lbs.

Figure 2-12 Units of measurement for concrete mixes

Chapter 3

Mixing, Placing, Curing, and Testing Concrete

To get the best and strongest concrete mixture, you should use the proportions of cement, sand, and gravel recommended by the experts who have tested the mixture you plan to use. These proportions are important if you want workable, durable, watertight, wear-resistant, strong, and economical concrete. You should also spot-check concrete that's delivered to you to be sure it consistently conforms to specifications.

A typical laboratory specification sheet will give you this information about a concrete mix:

- maximum aggregate size, in inches

- air content, as a percentage

- recommended maximum and minimum slump range, in inches

- exposure conditions (severe or mild) in air, fresh water, sea water, or sulfate (acid)

- maximum water/cement for exposure, as a percentage

- specified design strength, in psi

- water/cement ratio for strength

Measure the materials you use to make concrete within these limits:

- *cement* — plus or minus 1 percent

- *water* — plus or minus 1 percent

- *aggregate* — plus or minus 1 percent

- *admixtures* — plus or minus 3 percent

Each day you use them, check any admixture dispensers for accuracy. Using too much of an admixture can lead to serious problems in both fresh and hardened concrete.

Figure 3-1 shows typical concrete mixes for grades 1, 2, and 3 work. Grade 1 concrete work requires 2,000 to 3,000 psi structural concrete in foundation walls, footings, garden walls, and mass concrete. Don't use this concrete for work requiring abrasion resistance or watertightness. Grade 2 concrete work requires 3,500 to 4,000 psi concrete for driveways, patio slabs, watertight floors and walls, residential and barn floors, structural beams, columns, and slabs. Grade 3 concrete work requires the most strength: 4,200 to 5,000 psi. You should use this concrete for one-course industrial floors, thin sections, such as railings, ledges, posts, ornamental flower boxes, and building components with 1 inch or less concrete cover over reinforcement.

When you mix concrete at a job site, use this rule of thumb for 2,000 psi concrete: Use 1 part cement to 3 parts sand, 4 parts 1-inch rock, and $8^1/_2$ gallons or less of water. If you measure by cubic feet, for each cubic foot of cement, use 3 cubic feet of sand, 4 cubic feet of 1-inch rock, and $8^1/_2$ gallons of water.

Field Mixing of Concrete

Usually you use sacked cement as the basic unit for designing a concrete mix. One sack of cement is about 1 cubic foot.

Figure 3-2 shows the relative amount of ingredients in 1 cubic yard of stiff 4-inch-slump concrete and wet 8-inch-slump concrete. A stiff mix with large aggregate is about 15 percent water, 7 percent cement, and 78 percent aggregate. The aggregate has twice as much rock as sand. A wet mix with small aggregate is about 20 percent water, 14 percent cement, and 66 percent aggregate. The aggregate has equal amounts of rock and sand. Eight-inch-slump concrete has about 5 gallons more water, and will shrink more, than 4-inch-slump concrete.

Grade of work	Sacks of cement	Wet sand (lbs)	Gravel (lbs)	Wet sand (cubic ft)	Gravel (cubic ft)
1	5	1245	1935	$2^3/_4$	$4^1/_4$
2	6	1180	1915	$2^1/_4$	$3^1/_2$
3	7	1125	1895	$1^3/_4$	3

Figure 3-1 Typical concrete mixes

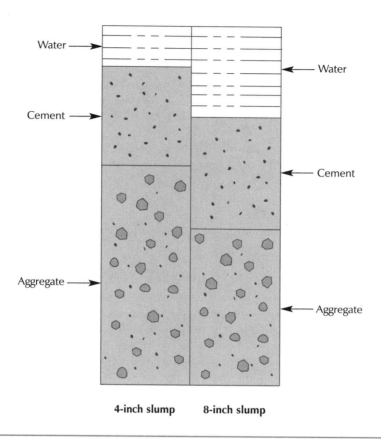

Water ——→

Cement ——→

Aggregate ——→

←—— Water

←—— Cement

←—— Aggregate

4-inch slump   **8-inch slump**

Figure 3-2 Relative amount of ingredients in 1 cubic yard of concrete

Cement paste (water and cement) is about 25 to 40 percent of the total volume of a concrete mix. Since aggregate is a large part of a mix, it's very important to use good aggregate. Use graded aggregate to make a strong mix that'll resist exposure conditions (freezing and thawing).

Specifying a Concrete Mix

There are several ways to specify a concrete mix. One is to order a particular strength of concrete, such as a 2,000 or 3,000 psi mix. The batch plant operators you order the concrete from will figure the quantities needed to get the strength you order. Another method is to order a mix by a particular water/cement ratio. This usually applies to concrete mixed at a job site. Figure 3-3 is a table with the water/cement ratio and the gallons of water you would use per sack of cement to get a particular strength of cement. The third method is to use a particular cement factor such as a 5-sack mix or a 6-sack mix.

Another approach is to use Abram's Law. This law uses a water/cement ratio by weight, i.e. pounds of water per pound of cement. To convert a water/cement ratio by weight to gallons of water per sack of cement, multiply the ratio by the

Water/cement ratio by weight	Water (gal. per sack)	28-day strength (psi)
0.45	5.0	5,000
0.49	5.5	4,500
0.53	6.0	4,000
0.57	6.5	3,500
0.62	7.0	3,000
0.66	7.5	2,500
0.71	8.0	2,000

Figure 3-3 Water/cement ratio for various concrete strengths

Compressive strength (psi)	Water/cement ratio by weight
2000	0.80
3000	0.69
4000	0.57
5000	0.47
6000	0.40

Figure 3-4 Compressive strength vs. water/cement ratio

weight of 1 sack of cement (94 lbs). Then divide your answer by the weight of 1 gallon of water (8.33 pounds) to get the number of gallons of water. Here's a formula for the calculation:

gallons of water per sack of cement =
(water/cement ratio × pounds per sack of cement) ÷ 8.33

Let's use the formula to find the gallons of water you need per sack of cement to make a mix with a water/cement ratio of 0.62:

gallons of water per sack of cement = (0.62 × 94) ÷ 8.33 = 7.0 gallons

Figure 3-4 shows compressive strength versus water/cement ratio. The water/cement ratio of a concrete mix will also vary depending on what size aggregate you use in the mix. For example, in 3,000 psi concrete, the water/cement ratio for 1^1/$_2$-inch rock is 0.62; for ^{3}/$_4$-inch aggregate, it's 0.69, and for ^{3}/$_8$-inch gravel, it's 0.70. Use 0.62 as an average water/cement ratio.

Storing Cement

You can store cement indefinitely as long as you don't let it come in contact with moisture. If you let cement absorb a small amount of moisture, it will set more slowly and its strength will be reduced. It's a good idea to store the sacks of cement in a place that's as airtight as possible. Stack the sacks on a raised floor to keep cement from absorbing dampness in the soil. Stack the sacks close together to reduce the amount of air circulation around them. Don't stack sacks along outside walls. For long-term storage, or where there's no shed to store cement in, cover the stacks with tarpaulins or other dampproof covering. To work out any lumps in a sack of cement, roll the sack on the floor before you open it. If you can't get the lumps out, don't use the sack for important work until you've had it tested by a laboratory.

On large jobs, bulk cement is usually transferred from a truck to elevated airtight and watertight bins. Concrete doesn't deteriorate; it loses effectiveness if it absorbs moisture from the air. If you use perfectly-sealed, airtight storage bins, cement will be usable indefinitely.

Weighing Aggregate

On fairly small jobs, you can weigh aggregate on platform scales that you set right on the ground. Build a runway to the scales so you can run a wheelbarrow onto one side of the scale, and off the other side. The amount of aggregate you place on each wheelbarrow load should be the same. Also, you should use an even number of wheelbarrow loads. So you'll need to figure out about how much aggregate you're going to use and divide it up into an even number of wheelbarrow loads. Doing this, you may not use full wheelbarrow loads. You can figure that a wheelbarrow holds about 2 to 3 cubic feet.

It's a good idea to build a measuring box to measure 1 cubic foot. Mark off the inside of the box in tenths of a cubic foot. You can also use a galvanized bucket as a measuring device. Most hold about $1/2$ cubic foot. It's a good idea to check the volume for sure, though.

Let's see how you would use a 3-cubic-foot wheelbarrow and a 1-cubic-foot box to make up the dry ingredients for a batch of concrete. Let's say you want a dry mix that's 1 part cement, 2 parts sand, and 3 parts coarse aggregate. Each batch will use 3 sacks of cement, so you'll need 6 parts sand and 9 parts aggregate.

- Use the 1-cubic-foot measuring box to load 3 cubic feet of sand into the wheelbarrow. Draw a line around the inside of the wheelbarrow at the level of the sand. Use 2 wheelbarrow loads filled to this level for each batch. Dump the 6 cubic feet of sand into the mixer's charge hopper. If the sand is damp, be careful measuring it because damp sand bulks up and takes up more volume.

- Measure the coarse aggregate in the wheelbarrow, using the 3 cubic foot level. Use 3 wheelbarrow loads of coarse aggregate (9 cubic feet) for each batch. Dump the rock into the charge hopper.

- Finally, add 3 sacks of cement directly into the hopper. Dump the contents of the charge hopper into the mixer. Measure and add the proper amount of water, using the required water/cement ratio, and mix.

Mixing Concrete by Hand

Even though field mixing is generally done by machine, you usually have to do some hand mixing. Use a clean surface for this. Build a wooden platform about 10 feet square out of T&G planks over a 2 × 4 frame. Make the joints between the planks tight to keep from losing mortar. Level the platform, then moisten the wood surface before you do any mixing. Place a measured quantity of sand on the platform. Spread the cement over the sand, then put the gravel on top. Use a hoe or a square-pointed D-handle shovel to mix the materials. Turn the mix at least three times until it's a uniform color. Add water slowly as you turn the mixture three more times. Add more water gradually until you get the consistency you need. It usually takes two workers to prepare a batch, one mixing and the other adding water.

Field Plant

When you need a large amount of concrete, or your job site is too far from a commercial batch plant, you may need a field plant. Field plants include central batch plants, batching and mixing plants, stationary plants, mobile field plants, portable field plants, prefabricated field plants, or a combination of these. A field plant usually includes single or multiple silos for storing aggregate and cement, front-end loaders, belt conveyors and dry or wet mixers.

Depending on the amount of concrete work you require on the project, you may erect a prefabricated batch and mix plant, subcontract the work, or lease a mobile or portable plant. You'll probably import raw material, like sand, crushed rock and cement, by hopper trucks or railroad cars. You can stockpile the fine and coarse aggregate in separate piles and move it with conveyors in a tunnel under the storage pile, or use front-end loaders. You can transfer the bulk cement from hopper trailers or railroad cars to storage bins with a crane or a pneumatic hose system. If there's insufficient local water, import water with tank trucks.

There are generally two types of field plants: central mix and transit mix. Most field plants are central mix, equipped with metering and transferring equipment that measures the required amount of cement, aggregate and water into the concrete mixer. After mixing, they load the fresh concrete into transit mix trucks that'll deliver the material to the work site. Central mixed concrete is a necessity when the concrete batch plant is too far away from the job site.

In a transit mix plant, the preweighed material is loaded dry into a truck and delivered to the job site, where water is added and the concrete is mixed. You have more control over a mixture if you get it from a transit mix plant.

Transit Mix Concrete

Concrete is usually delivered to a job site by transit mix truck. If you get your concrete this way you don't have to store concrete materials, and you don't have to batch or mix it. The trucks will bring you concrete ready for the forms. Each truck has a rotating drum, water tank, and measuring devices mounted on the truck chassis.

Agitator trucks are similar to mixer trucks but they don't have water tanks. The dry cement, aggregate, admixtures and water are loaded into the truck at the plant and mixed as the truck travels to the job site. The advantage of agitator trucks is that they operate from a central mixing plant where quality concrete is produced under controlled conditions. Discharge from agitators produces a uniform, homogeneous concrete mix. Timing of the delivery must suit the job organization and the concrete crew and equipment must be on site and ready to handle the concrete when the truck arrives. Mixing and operation of the truck-mixer and non-agitating units should conform to ASTM C-94, Specifications for Ready-mix Concrete.

Another concrete mixing procedure is called *shrink mixing*. The material is agitated in a stationary mixer at the plant just enough to lightly mix the ingredients. Then, mixing is completed while the mixer truck travels to the site. The shrink method reduces the amount of time the concrete must be mixed in transit. This method gives you better control over a mixture.

Usually, you get a load ticket with each batch of premixed concrete delivered to you at the job site. Here's an example of what you'll usually find on a load ticket:

- *amount of cement:* 4,230 pounds

- *amount of sand:* 9825 pounds

- *amount of No. 2 gravel:* 5,850 pounds

- *amount of No. 3 gravel:* 7,560 pounds

- *amount of No. 4 gravel:* 1,719 pounds

- *amount of water added at plant:* 180 gallons

- *total allowed water:* 270 gallons

- *free moisture:* 66.8 pounds

Figure 3-5 shows how many square feet 1 cubic yard of concrete will cover for various slab thicknesses. You can use this table to estimate the amount of concrete you should order for a large concrete slab. Since a cubic yard is equal to 27 (3 × 3 × 3) cubic feet, a 4-inch slab will use 1 cubic yard of concrete for each 81 (27 × 3) square feet of area in the slab. By way of an example, let's say you have to place 10,000 square feet of 6-inch-thick warehouse floor. You'll need 185.2 (10,000 ÷ 54)

Slab thickness, inches	Area, square feet per cubic yard of concrete
2	162
3	106
4	81
5	65
6	54
7	46
8	40
9	36
10	32
11	30
12	27

Figure 3-5 Area covered by 1 cubic yard of concrete

cubic yards of concrete. Assuming a mixer truck carries 10 cubic yards, the math gives you 18.52 truckloads. You'll need a little extra for waste, so order 20 truckloads.

Mobile Mixing Units

Truck or transit mixers mix the material by rotating an agitator in one direction, and discharge the material by rotating in the opposite direction. The blades tend to screw the concrete mixture back to the opening. Mixing speed is between 4 and 12 rpm. Agitating is slower. Mixing requires between 70 and 100 rotations of the drum but not more than 250. The mixing time for a 6-cubic-yard mixer is between 3 and 10 minutes.

For further information, contact the National Ready-mixed Concrete Association (NRMCA), the Truck Mixer Manufacturer Bureau (TMMB), or the Concrete Plant Manufacturing Bureau (CPMB).

Working Out Delivery Specifications for Ready-Mix Concrete

When you order concrete from a ready-mix supplier, remember that mixers come in many standard sizes. Usually a mixer will hold from 4 to 12 cubic yards. Give the supplier a description of the mix you need by number or other identification. Tell the supplier the total cubic yards you want. Specify the rate of delivery in cubic yards per hour, or by the spacing of the trucks. Be sure to allow enough lead time. The supplier will want to know where to deliver the concrete mix and any unusual delivery conditions. Describe how you want the concrete to be placed.

There are three things you should be concerned with when you have cement delivered from a concrete mixing plant to the job site. They are:

1. Make sure the concrete is delivered to you in good time so it doesn't dry out or lose its workability or plasticity before it's placed.

2. Make sure aggregates and paste are separated as little as possible to help keep the concrete uniform. Try to prevent any fine material, cement, or water from being lost.

3. Organize delivery so there's no delay in placing concrete for any particular unit or section so you don't have any undesirable fill planes or construction joints.

Placing Concrete

Here's a checklist you can review before you pour any concrete:

■ Make sure any excavation you did for the foundation is dry and free of water, snow, or ice.

■ Don't place concrete on porous ground.

■ Sprinkle or seal semiporous subgrade to prevent suction that could make a mix too dry and stiff. Just make dry soil slightly moist so that it won't draw water from the fresh concrete. Otherwise, the concrete may get too stiff to work or place properly.

■ Make sure you've put up stable and complete formwork.

■ Make sure you've put the reinforcing steel and other imbedded items, such as expansion joints and anchor bolts, securely in place.

■ Set the openings and sleeves for ducts, pipes and conduits and have them checked by the job superintendent.

■ Have the forms and reinforcing bars inspected and approved by the building inspector.

■ Remove and clean all dirt from contact surfaces so that the fresh concrete will bond well. Clean the top of footings before you pour the pedestals, and brush off all construction joints in hardened concrete slabs and walls before you extend the slab or wall with fresh concrete.

- Make sure all equipment is clean. You should clean it at the end of each operation or day's work.

- Remove any hardened concrete and foreign materials from the conveying equipment.

Here are some rules to follow as you pour concrete:

- Mix the concrete thoroughly as you pour it. You can use a rod, but a mechanical vibrator works better. High frequency vibration will keep the coarse and fine aggregate from separating. You can use an electric-powered, pneumatic, or gasoline vibrator.

- Don't dump concrete directly from a mixer into a bucket. This throws the heavier and larger rock to one side.

- Place concrete continuously in layers.

- If you use a buggy (a cart used to carry fresh concrete) to pour concrete into a formed wall, unload it in a concentric fashion using a vertical chute. Figure 3-6 shows a buggy, metal chute, and down pipe used to pour concrete into a wall form.

- Pour the concrete in even horizontal layers, 6 to 24 inches in depth.

- Be sure to place each new layer before the layer below it sets up.

- Pour concrete as close as possible to its final position.

- Don't place concrete in large quantities at one point and let it run or be worked a long distance in the forms. This may segregate the material because the mortar (cement, sand, and water) will tend to flow out ahead of the coarser aggregate.

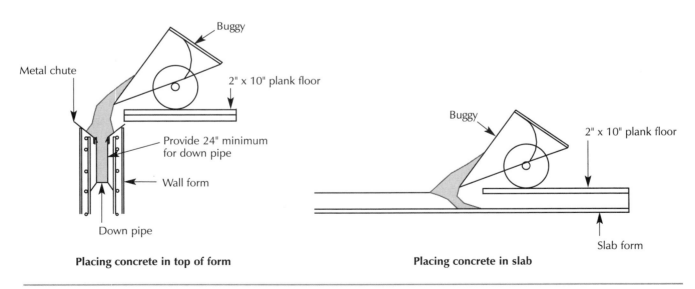

Placing concrete in top of form   **Placing concrete in slab**

Figure 3-6 *Pouring concrete in forms*

- Place concrete in horizontal layers of uniform thickness. Make each layer 6 to 12 inches thick for reinforced members, and up to 18 inches thick for mass concrete work. Be sure to thoroughly vibrate each layer before you place the next one. That removes air pockets and fills voids between the sand or rock particles, which makes the mix denser and more compact. Vibrating floor slabs also pushes larger size aggregate lower so it doesn't interfere with the screeding and finishing operation.

- Pour concrete at the far end of slabs, against the concrete you just placed.

- Avoid bouncing a mixture off one side of a form. This makes heavier aggregate bounce farther than fine aggregate, so the mixture separates.

- If you can't get a mixer truck up to your forms, pump concrete from the truck to the forms.

- Dump concrete from a transit-mix truck by chute into a hopper. Then pump the mix through a hose mounted on a specially-designed truck with an articulated boom.

- Don't let concrete drop freely more than 3 or 4 feet. In thin sections, use rubber or metal drop chutes. In tall, narrow forms, you can place the concrete through openings, or windows, in the sides of the forms. Don't let a chute discharge directly through an opening as this may cause the mix to separate. Use a rectangular chute with a steel hopper at the top for placing concrete in narrow forms.

- Keep fresh concrete from bleeding by using low-slump, air-entrained concrete that has adequate cement and properly graded sand. Also, don't float or trowel concrete until it's hardened enough so the water and fine material in it don't come up to the surface.

- Don't spread dry cement on a wet surface to take up excess water. That can form a paste that will break off, or cause dusting and scaling. It's better to wait for all water to evaporate or be removed. You can drag a rubber garden hose over the surface to wipe the water off. Another way is to use fans or blower type heaters to dry off the water.

Some means you can use to place concrete are:

- crane and bucket: a bucket holds about $1/2$ to 4 cubic yards.

- floor hoppers with single and double gates at the hopper bottom: a hopper holds about 1 to 10 cubic yards.

- forklift buckets carried by forklift trucks

- belt conveyors: a mobile unit can pour as much as 100 cubic yards of concrete per hour, up to three stories above or below grade.

- pneumatically applied concrete (also called Gunite and shotcrete): a high pressure pump can send out a very dry mix that makes dense strong concrete.

- telescopic boom truck: can put out as much as 80 cubic yards of concrete per hour, 320 feet vertically or 1250 feet horizontally. Use this when you need to pour concrete for the upper floors of a high-rise building.

- power-operated buggies

- general purpose concrete bucket

- lay-down concrete bucket

Placing Concrete with a Conveyor

Concrete has been placed by conveyor for many years to get around or over obstructions that are hard to overcome with other methods. Some conveyors are self-propelled and can be attached under the discharge chute of a truck mixer. Because these units can reach over walls, ditches, and other obstacles, they save time and the cost of rehandling concrete. Belts can run at speeds up to 900 feet per minute, and concrete can be placed at the rate of 300 cubic yards per hour.

If you set up the conveyors in series, with the discharge end overlapping the next conveyor, they can carry a continuous flow of concrete over a great distance. You can start placing concrete at either end of a job and extend or retract the belt system to continue placing concrete without interrupting the pour. You can also use a side discharge arrangement so that you can place concrete at any point along the conveyor. This allows placing concrete over a wider area.

Use a rubber scraper to prevent loss of concrete on the return belt. Use covers over a long belt to protect the concrete from rain and hot sun.

Defects Caused by Faulty Placement

Here are some common defects in concrete caused by faulty placement:

- Foundation settled or cracked because it was placed on poor, uncompacted, or expansive soil.

- Concrete spalled because there wasn't enough concrete over the steel reinforcement in the concrete. When moisture can reach reinforcing bars they rust, expand, and crack the concrete they're in.

- Concrete damaged by alkali and acid because it didn't have enough thickness to cover the nearest steel reinforcement bar. If alkaline or acidic water penetrates the concrete and reaches the steel reinforcement, the bars may corrode and expand, which results in concrete spalling.

Always try to keep any concrete mixture from separating. Take special care with certain mixes that have a tendency to separate, such as harsh mixes, mixtures that are very wet or very dry, and mixtures with very little sand. A harsh concrete mix is one that doesn't have enough sand in it to fill in the spaces around the coarse aggregate in the mix. The surface of a harsh concrete mix tends to be rough.

Once a concrete mixture has separated, you can't make it right again. To avoid this, don't dump concrete into deep forms without using a chute. Don't move concrete a long distance horizontally. Be sure to tamp and vibrate fresh concrete.

Always try to keep concrete slabs from bleeding. This can happen when you screed or level off concrete. The surplus water (also called *water gain*) in the concrete rises to the surface. The fine aggregate in the mix also rises to the surface making voids and cracks around the coarse aggregate and the reinforcement in the concrete. Use dry, sandy subgrade soil to absorb part of the water in a mix and reduce bleeding. Using impervious wet subgrade soil will increase bleeding.

That gray, nearly white, substance formed on the surface of some concrete is called laitance. It's water, cement, and fine aggregate rising to the surface. Laitance is caused by adding too much water to a mix, or vibrating, troweling, or floating the mix too much. Laitance is bad for horizontal joints as it reduces bonding.

Placing Cellular Lightweight Concrete

Cellular lightweight concrete, the kind you use as underlayment in a multistory building, is pumped or otherwise transported to a job site. Then you screed the concrete to the proper thickness, usually between $1^1/_2$ and $1^5/_8$ inches. Use a darby and trowel to produce a flat surface. Keep load and foot traffic off the concrete for at least 24 hours.

Lightweight concrete and cellular lightweight concrete are different. Lightweight concrete uses a lightweight aggregate such as expanded shale, pumice, or expanded slag. Cellular lightweight concrete contains millions of air bubbles. The bubbles are produced by a special admixture in the concrete that foams up and makes the mixture swell.

Making a Cellular Lightweight Concrete Test Cylinder

You should take specimens of a cellular lightweight concrete mixture at the point of placement and put them in 3 × 6-inch cylinders. Protect the test cylinders from weather and disturbance after you cast them. After 24 hours, move the cylinders to a curing room. Cure the specimens with moisture for 7 days, then dry cure them for another 21 days, or a total of 28 days. Then test the cylinders.

Here's how to make a cellular lightweight concrete test cylinder:

1. Take a sample from the mix you're testing within 15 minutes of making the mix.

2. Moisten the test mold and equipment.

3. Remix the sample and put it in the mold in three equal layers.

4. Distribute the mix evenly when you fill the mold before rodding it.

5. Consolidate each layer by making 25 strokes with a tamping rod. Distribute strokes uniformly over the mold, going about 1 inch into the previous layer.

6. Tap the sides of the mold lightly to close voids left from rodding.

7. Strike off the top surface with a tamping rod and float or trowel the surface.

8. Clean off the outside of the mold.

9. Put an identifying mark on the mold to identify it as a test mold.

10. Cover the mold with oiled glass, steel plate, a plastic bag, or plastic cap to keep the sample from drying out.

11. Cure the sample in the mold for the first 16 to 48 hours at an air temperature of 60 to 80 degrees F.

12. Remove the sample from the mold after 16 to 48 hours and store (final cure), in moist condition, at 73.4 degrees $+/-3$ degrees F.

Later in the chapter there's a section on testing that test cylinder.

Pumping Concrete

A common type of concrete pump is mounted on a trailer. It's hydraulically operated, and pumps concrete through pipes and hoses. You can pump concrete through a 5- to 8-inch steel pipeline to locations that you can't get a buggy into. Depending on the size of the pump and pipe, you can pump concrete 600 to 1,000 feet horizontally. Reduce this distance by 40 feet for each 90-degree bend, and 20 feet for each 45-degree bend. You can also pump 1 foot vertically for every 8 feet you pump horizontally. You can pump 20 to 65 cubic yards of concrete per hour if the aggregate in the mix is 3 inches or less in size. For successful operation, feed a constant supply of concrete to a pump. It's also a good idea to use an agitator to remix concrete as it's dumped into the hopper.

Here are some recommendations for pumping concrete safely:

■ Keep the hood closed whenever you operate the pump.

- Don't use damaged or worn couplings. Inspect hose fittings regularly to make sure coupling grooves aren't worn down.

- Clean concrete pump, pipes and hoses with water, or compressed air under certain conditions. For cleaning the hose, use a plug of material compressed to the same size as the pipeline. The plug, called a *go devil*, may be of sponge, burlap, or wadded cement bags. Insert the plug into the distribution line and drive it with water or air pressure. Water is safer. Compressed air should only be used in tunnels or down grade runs to foundations where it's difficult to dispose of water. For safety, attach a "devil catcher" or trap basket at the end of the line. If you don't, that plug will become a dangerous projectile.

- Before you discharge any concrete into a hopper, spray 3 to 4 gallons of water into the hopper, followed by 5 gallons of creamy cement and water slurry ($1/2$ bag of cement to 5 gallons water). This will lubricate the hose and help keep it from getting blocked up. It's very important that you get the concrete to flow properly at the start of the pumping cycle.

If you have to stop a pump:

- Leave the hopper full of concrete.

- After 3 minutes, turn off the engine. Vibrating the mixture in the hopper can make it separate and block the manifold.

- After 10 minutes or more, start the engine and pump 6 or 8 strokes every 5 minutes to keep the concrete from setting up.

- If you have to wait an extended period of time, wash out the pump and hoses and start over.

- Don't try to force a pump to clear a blocked hose or you may damage the pump. Find out where the blockage is and clear it before you do any more pumping.

Spraying Concrete (Gunite or Shotcrete)

Use spray-applied concrete to stabilize rock slopes, make thin slabs, or repair concrete work. To use this material, you combine it in a mixer and then place it in a two-compartment vessel, called a gun. The gun will discharge a steady stream of concrete. When you apply the stream overhead, you can lose up to 50 percent of the mix. When you point the gun downward, you can lose 15 percent of the mixture due to rebound. Apply concrete in $3/4$-inch-thick layers on vertical or overhanging surfaces and up to 3-inch-thick layers on horizontal surfaces.

Underwater Concrete

To make concrete for underwater placement, use a 7-sack mix with 50 percent sand and 50 percent $1^1/_2$ to 2-inch aggregate. The mix should have a 6-inch slump. That's the easy part — placing concrete underwater is a whole different ballgame. It's always best to place concrete in air rather than under water. If you have to place concrete in water and you're not sure about your own expertise with such a job, get an experienced person to supervise the work. Then follow these steps:

- Set the forms on the bed to maintain a seal.
- Bury the end of the discharge hose in the freshly placed concrete.
- Don't place concrete if the water temperature is under 45 degrees F.
- Don't place concrete if the water velocity is over 10 feet per minute unless you use sacked concrete or a cofferdam.

If the water you're working in is moving less than 10 feet per minute, you're okay. However, if the current is flowing faster than that, you should build a cofferdam to slow down the flow. A cofferdam is a watertight enclosure you build around the forms so that you can pump water out from around the forms. You can build a cofferdam by driving interlocking steel sheet piling or overlapping wood piling into the ground. Water may seep into the work area through the soil or cofferdam. Wait 24 hours to allow freshly placed concrete to set before you pump out the "seeped" water.

You can place concrete at a considerable depth below the water surface if you use open-top buckets with a drop bottom. Use slightly stiffer concrete than you use with a tremie, but the mix should always contain 7 sacks of cement per cubic yard of concrete. Fill the bucket completely and cover the top with a canvas flap. Lower the bucket slowly into the water so the canvas won't be displaced. Don't discharge the concrete from the bucket before the bucket reaches the bed's surface. Take soundings frequently to make sure you're keeping the top surface of the concrete level.

Using a Tremie to Place Concrete Underwater

A tremie is a pipe with a funnel at its upper end, and a gate at its lower end. See Figure 3-7. You close the gate at the lower end, and then fill the pipe by adding concrete at the funnel end before you put the pipe in water. It's best to pump concrete into the funnel at the top of the tremie. You can also use open-top buckets with a drop bottom to fill a tremie.

To place concrete, open the gate from above and lift the tremie slowly to let the concrete flow. Be sure to keep the bottom of the pipe continuously buried in the newly-placed concrete. If the bottom of the pipe comes out of the concrete, raise the tremie, plug the lower end, and lower the tremie in position again.

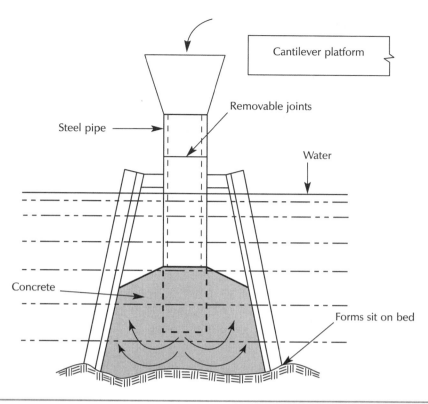

Figure 3-7 *Placing concrete underwater with a tremie*

Don't let any air or water get into the pipe. Keep it constantly filled with concrete. Also, don't move a tremie horizontally. Always lift it and move it to a new position.

Keep the top of the concrete as level as possible. If you're doing large pours, use a number of tremies.

Curing Concrete

After you have poured a batch of concrete, the next important phase is curing, or setting. Curing concrete takes time. Concrete cures as long as there's water in it. The curing process stops when the concrete is dry. Normally, concrete gains most of its full strength in 28 days. It will continue to get stronger after that, but not a lot. Figure 3-8 shows normal concrete strength versus curing time.

Concrete that dries out too soon won't develop its full strength. For maximum strength concrete, keep it moist during the entire curing period. In hot, dry weather, cover fresh concrete with wet burlap or straw. You can also spray a fresh

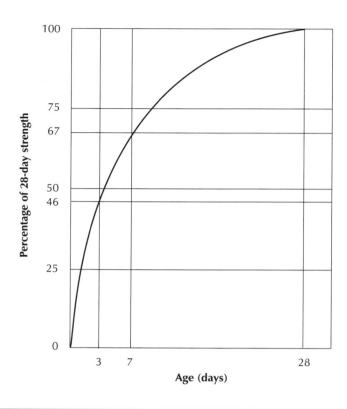

Figure 3-8 *Normal concrete strength vs. curing time*

concrete surface with a curing compound made of a blend of oils, resins, waxes, and solvents. When the solvent evaporates, it leaves a membrane on the surface that seals in the original mixing water.

If concrete dries too soon, spray its surface with water. The chemical reaction will continue, but the concrete won't be as strong as if you had kept it moist all the time. The longer a mixture is left dry, the weaker it becomes.

You can use plastic sheets (polyethylene and reinforced polyethylene film) to help cure and protect concrete slabs. They don't mold, mildew, or become brittle if exposed to sunlight. Usually you can get them in 8-foot-wide sheets, 125 feet long. The film is 4 to 10 mil thick, in clear or black. In subfreezing weather, you can use electric blankets to protect freshly-poured concrete. You can also use these blankets to accelerate the curing process.

The outside temperature will affect how concrete cures and how strong it'll be. Concrete cures faster when the weather is warm than when it's cold. Below 40 degrees F, it will set slowly. The best temperature range for pouring concrete is between 50 and 70 degrees F. At higher temperatures, the strength increases in the first few days but is lower in later periods. At lower temperatures, the strength is lower at all stages.

Testing

Bad test results don't always mean bad concrete. Sometimes, an owner blames the contractor for poor concrete work. Or, the contractor blames the ready-mix producer for supplying poor quality concrete. On the other hand, the ready-mix producer may blame the laboratory for faulty testing, or the laboratory blames the contractor for improperly handling the test specimens. If you have a problem, here's what you should do:

1. Verify the accuracy of the tests run on the concrete. Low strength results may be due to:

   ■ improper concrete sampling

   ■ improper specimen preparation

   ■ improper field curing

   ■ improper laboratory curing

   ■ improper capping of specimen

   ■ improper testing procedure

2. Compare the structural requirements with the measured strength. The concrete may still be strong enough for the anticipated loads.

3. Perform nondestructive tests such as the Schmidt Impact Hammer Test.

4. Make core tests by drilling cores and subjecting them to compression tests.

5. Make load tests with drums of water or hydraulic jacks.

6. Perform corrective measures such as installing steel columns or beams under questionable members.

In the case of an improperly-constructed concrete floor slab, the building department may condemn the building until it's proven safe. If calculations indicate that the slab fails to meet the required strength, the building department may permit the owner to make load tests on the completed structure. If the load tests meet the code requirements, the building department may accept the building.

Testing the Strength of a Batch of Mortar

You can make samples of mortar that a laboratory can test for compressive strength. You need to prepare three 2-inch cubes of the mortar with the proportions of cement, sand, and water called for in ASTM Specification C270.

Another type of test for mortar is specified in *UBC* Standard 24-22. To do this test:

- Spread $1/2$- to $3/8$-inch-thick mortar on a concrete block and let it stand for 1 minute.

- Then remove the mortar and place it in a 2-inch by 4-inch cylindrical mold in two layers, compressing it into the mold with a flat-end stick, or your fingers.

- Lightly tap the mold on opposite sides, then level it off.

- Immediately cover the mold and keep it damp until you take it to the laboratory.

- After 48 hours, the laboratory will remove the mold and place the sample in a fog room for 28 days until it's tested in damp conditions.

Mortar should have a minimum compressive strength of 1,500 psi.

Testing the Strength of a Batch of Concrete

To test the strength of a batch of concrete, pour a sample of fresh mixture into a cylinder 6 inches in diameter and 12 inches long. Usual practice is to make enough test cylinders so you can test the concrete after it's cured for 3, 7, and 28 days. Let test cylinders cure at least 3 days.

You can test a sample using a machine that applies weight to the sample until it breaks. Record the force (in pounds) it takes to break each sample. Divide the force by the cross-sectional area of the cylinder (in square inches) to find the compressive strength in pounds per square inch, such as 2,000 psi or 3,000 psi. The cross-sectional area of a 6-inch diameter cylinder is 3.14 times the diameter squared and divided by 4. Here's how to figure that out:

cross sectional area = (3.14 × 6 × 6) ÷ 4 = 28.26 square inches

Engineers often use the term *kip* in place of 1,000 pounds. You can also use the word *pound* instead of psi to tell the compressive strength of concrete. For example, 2,000-pound concrete has a strength of 2,000 psi.

Test Loading Floor Slabs

If test cylinders weren't taken or the test results were below required strength, you can resort to test loading of the subject concrete roof or floor slab. Check the existing elevation of the top of the slab near the supporting columns, because there's no deflection at these points caused by the weight of the slab (unless the columns and foundations have settled).

Here's how to make a load test on a slab:

- Use a transit or builder's level to find out the elevation of the edges and center of the slab before you put a test load on it. The elevation over a column is usually used as the datum (benchmark).

- Figure out the dead and live loads that the slab will have to bear. The dead load is the weight of the slab itself, based on a cubic foot of concrete weighing 150 pounds. Therefore, the dead load of a 6-inch thick slab is 75 psf. The live load is the superimposed load that the slab is designed to carry. For example, a live load 40 psf is required for a residential floor, and 100 psf for a storage room.

- Figure out the test load for the slab. This is 85 percent of the sum of 140 percent of the dead load and 170 percent of the live load.

- Apply the test load in four equal increments. For example, a test load of 204 pounds per square foot would be applied in four increments of 51 pounds per square foot each. You can use containers of water or sandbags to apply a test load. Place the load increments over the entire area of the slab you want to test.

- After the full test load is on for 24 hours, use the transit, or builder's level, and a leveling rod, to measure the elevation at the edges and center of the slab again.

- Remove the test load.

- If there are cracks in the slab, it has failed.

At this point, if there are no cracks in the slab, you need to figure out how much the slab has changed since you put the test load on it. The slab shouldn't change more than $(l \times l) \div (2{,}000 \times h)$, where l is the width of the slab in inches and h is its depth in inches.

A slab or beam deflects when it's loaded. When the load is removed, the slab or beam returns to its original position. This is called *rebound*, or *elastic deformation*. If the load is greater than the slab or beam is designed for, a permanent deflection occurs when it's removed. This is called *plastic deflection* or *deformation*. After 24 hours, the slab should rebound by at least 75 percent of that amount. If it doesn't, you can test the slab again. If the next test is still unacceptable, you'll have to replace it or reinforce it.

Let's look at an example for a slab that's 20 feet wide (240 inches) and 6 inches thick. A square or nearly square slab is designed to span in two directions. A rectangular slab is designed to span in one direction, which is usually the shorter direction. Let's say the dead load on the slab is 50 psf and the live load is 100 psf.

$$\text{test load} = 0.85 \, (1.4 \times 50 + 1.7 \times 100)$$

$$= 0.85 \, (70 + 170)$$

$$= 204 \text{ psf}$$

The deflection is measured at midspan, or the center of the slab, where the maximum sag occurs. The maximum amount the slab can change is (240 × 240) ÷ (2,000 × 6) or 57,000 ÷ 12,000, or 4.8 inches. Therefore, if the maximum change is less than 4.8 inches with a full load, the slab is acceptable.

If the maximum change is 5 inches but the slab recovers more than 3.75 inches 24 hours after you remove the test load, the slab is acceptable. If it doesn't recover the 3.75 inches, test it again. If it still fails, reinforce or replace it.

Testing Freshly-Mixed Air-Entrained Concrete

The advantages of air-entrained concrete are:

- It increases the freeze-thaw resistance 10 to 20 times over regular concrete.
- It improves resistance to salt-action and scaling.
- It improves resistance to sulfate action.
- It improves the watertightness of the concrete.

There are three air-entrainment test methods: pressure method, volumetric method, and air indicator method. You should get a licensed laboratory to do these tests for you. Building departments usually require a compression strength test for concrete that has to be more than 2,000 psi. Here's what the 1997 *Uniform Building Code* (Section 1905.6) has to say on testing concrete:

Samples for strength tests for each class of concrete placed each day shall be taken not less than once a day, nor less than once for each 150 cubic yards of concrete, nor less than once for each 5,000 square feet of surface for slabs and walls.

On a given project, if the total volume of concrete is such that the frequency of testing required would provide less than five strength tests for a given class of concrete, tests shall be made from at least five randomly selected batches or from each batch if fewer than five batches are used.

A strength test shall be the average of the strengths of two cylinders made from the same sample of concrete and tested after curing for 28 days.

If a compression strength test fails, there are several tests the building department may require. They may want you to take three more cores from the hardened concrete and get them tested. Also, you may have to get a physical load test made on the structure. This test involves loading the suspect floor slab with barrels of water. If all the tests fail, the building department could make you remove all defective concrete work.

Testing Aggregate

The standard test for grading fine and coarse aggregate is described in ASTM C-33. For fine aggregate, you stack a set of sieves, one above the other. Usual sieve sizes are #4, #8, #16, #30, and #50. For coarse aggregate, use 6 inch, 3 inch, $1^1/_2$ inch, $^3/_4$ inch, $^3/_8$ inch, and #4 sieves. These sieve sizes are based on square openings.

Nondestructive Testing Devices

Using a nondestructive testing device at a job site has many advantages. You don't have to make up and deliver specimens to the laboratory and wait for their results. And you get the test information immediately so you don't have to delay the progress of the job. However, you have to keep all instruments properly maintained and calibrated. The manufacturer of the instrument will normally provide manuals and how-to-use training for the instrument.

You can use a number of portable instruments to test the strength of hardened concrete, locate and size embedded reinforcing bars, measure crack widths and depths, or determine a wall's watertightness. James Instruments, Inc. makes the following instruments for nondestructive testing of hardened concrete:

■ Rebound Hammer conforms to ASTM C7311 and C805, and indicates compressive strength by a spring-loaded activated percussion weight creating an impact and rebound effect.

■ Calibration Hammer, conforms to ASTM C7312, also indicates compressive strength.

■ Windsor Penetrometer, which conforms to ASTM C803, measures compressive strength in concrete. It's a self-powered device that drives a probe-sleeve against concrete to be tested and measures the resistance to penetration.

■ Windsor Probe conforms to ASTM C803, and measures compressive strength in concrete by an explosive charge that drives a steel probe into the concrete.

■ HR Rebar Locator finds position, depth and size of reinforcement bars. This device indicates bar size to an 8-inch depth and location to a 12-inch depth. See Figure 3-9.

■ Rebar Datascan locates position, depth, and size of reinforcement bar.

You can also use these James instruments for testing new or fresh concrete:

■ PPR Meter measures compressive strength in new concrete by pin penetration.

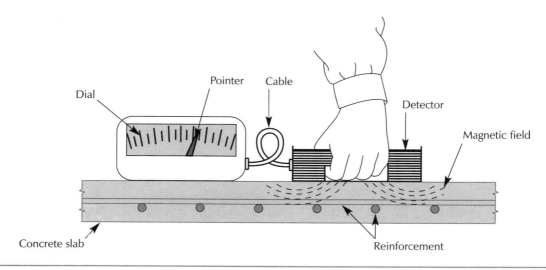

Figure 3-9 *How to use an R-meter — reinforcing bar locator and sizer*

■ Pullout Test System measures compressive strength of fresh concrete and tensile strength of mature concrete by a hydraulic ram, a control valve, and gauges. It measures the force to displace an embedded steel disk in concrete.

■ M-Meter Maturity System measures developing strength and predicts the later strength of concrete.

■ RT-Meter monitors the temperature of maturing concrete.

You can use Germann Instruments, Inc.'s and other manufacturer's devices for nondestructive testing hardened concrete:

■ Lok-Test finds the compressive strength near the concrete surface.

■ Capo-Test also finds the compressive strength near the surface.

■ DocTer measures thickness, delaminations, honeycombing, voids, depths of surface openings and cracks.

■ CMD, Crackscope measures crack widths and deposits at crack edges and shear opening displacement of a live crack.

■ Moisture Detector evaluates the watertightness, surface cracking, porosity, and water permeation.

Destructive Testing Devices

You can use an electric-powered coring drill to cut cores from concrete for testing. Some drills are hand-held. Others must be mounted on the floor or supported from a wall. Some of the manufacturers of these instruments are James Instruments, Germann Instruments, Inc., and Mitchell Instruments Co.

Chapter 4

Surveying

Field surveying, sometimes called construction surveying, is an extremely important part of concrete work. It begins when the land surveyor, who is licensed by the state, has finished surveying the property lines. A construction surveyor doesn't have to be licensed, but he does have to know how to operate and maintain survey equipment. Also, he must understand trigonometry to accurately lay out the foundation forms as called for on a set of plans. He's responsible for checking all the dimensions shown on the plot and foundation plans and notifying the job architect or engineer if a string of dimensions doesn't add up.

This is your last chance to check the plans for mathematical errors. If you put a foundation in the wrong place, even though the error might not be yours, don't be surprised if somehow it manages to come down on your head. Removing a foundation built in the wrong place is one of most costly things that can happen on a job. When there's this much money involved, honesty and decency may take a back seat! So don't take any dimensions shown on the plans for granted. Add up each string of dimensions yourself to be sure they equal the total lengths shown on the plans. Double check the foundation elevations. If you think there may be an error, don't even think about going ahead with the job. Notify the architect or engineer immediately, in writing.

Here are some of the tasks a construction surveyor will do:

■ Set up transit lines, coordinate lines and job bench marks needed to lay out foundations, concrete structures and roads.

■ Control horizontal distances and elevations in formwork.

■ Make a topographic survey for grading, excavation and fill.

- Make a quantity survey for earthwork.

- Prepare profiles and cross sections for road construction.

- Prepare as-built plans of site work. As-built plans are drawings that show all changes made since the building department approved the plans. These include field changes, revisions, and change orders relating to the building or site work. The contractor, subcontractor, or surveyor indicates these changes in ink on the approved plans.

Plot Plans

A plot plan is a drawing of a property that includes the building site dimensions, location of the building and property boundaries, utility services, etc. Plot plans are usually drawn to an engineering scale. On a plot plan, a foot is divided into tenths of a foot. On a foundation plan, a foot is divided into inches. So be prepared to have a chart ready to convert from one scale to the other. Figure 4-1 is a sample conversion chart.

To begin to use a plot plan, study it to see how the land surveyor marked the property corners, and then locate a datum elevation. A datum elevation is a point marked on a permanent feature, such as a concrete curb or sidewalk, that you can use as the arbitrary bench mark for a project. Usually you assign an elevation of 100.00 to this point. But you always need to show its true elevation in reference to the public street elevation. For example, you can say your datum elevation of 100.00 is equal to USGS Elevation 345.67. USGS stands for United States Geological Survey.

Fraction of an inch	Decimal of an inch	Decimal of a foot	Fraction of an inch	Decimal of an inch	Decimal of a foot	Fraction of an inch	Decimal of an inch	Decimal of a foot
$^1/_{16}$	.0625	.0052	$^3/_8$	.3750	.0313	$^{11}/_{16}$	.6875	.0573
$^1/_8$	.1250	.0104	$^7/_{16}$	.4375	.0365	$^3/_4$	.7500	.0625
$^3/_{16}$	.1875	.0156	$^1/_2$	.5000	.0417	$^{13}/_{16}$	.8125	.0677
$^1/_4$	.2500	.0208	$^9/_{16}$	.5625	.0469	$^7/_8$	.8725	.0729
$^5/_{16}$	.3125	.0260	$^5/_8$	.6250	.0521	$^{15}/_{16}$	.9375	.0781

Figure 4-1 *Converting fractions of an inch to decimals of an inch and foot*

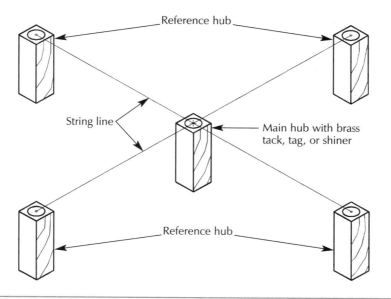

Reference hub

String line

Main hub with brass
tack, tag, or shiner

Reference hub

Figure 4-2 *Reference stakes*

Set extra reference stakes around important points so you don't have to get a second survey if the main control marks are disturbed during construction. Figure 4-2 shows reference stakes. Highlight these stakes with brightly colored cloth or plastic strips, called flagging. Also, use white paint to make the wood hubs stand out, or use shiny metal disks, called shiners. It's important to record in a field logbook the location of all stakes you set.

You'll have to use some trigonometry for much of your field layout. The most important rules of trigonometry are shown in Figure 4-3.

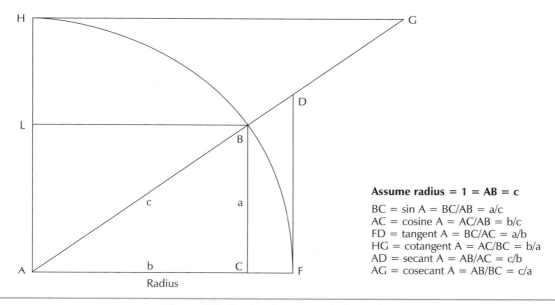

Assume radius = 1 = AB = c

BC = sin A = BC/AB = a/c
AC = cosine A = AC/AB = b/c
FD = tangent A = BC/AC = a/b
HG = cotangent A = AC/BC = b/a
AD = secant A = AB/AC = c/b
AG = cosecant A = AB/BC = c/a

Figure 4-3 *Basic trigonometric formulas*

Transits

Although most public works and large construction projects are laid out with electronic instruments, the traditional transit is still the basic tool for medium and small jobs. Surveyors use a transit, or a level and a leveling rod, to measure the relative difference in elevation of any point and the datum elevation. Mostly, you'll use a transit to set horizontal and vertical angles. If you're only setting elevations, it's more effective to use a builder's level. Its telescope has greater magnifying power than a transit's. Also, it has fewer moving parts than a transit, so it's easier to set up and keep steady.

Figure 4-4 shows a standard precise transit. The horizontal and vertical circles are divided into degrees and 30-minute increments ($^1/_2$ of a degree). You classify a surveyor's transit by the smallest angle that you can read on the transit's vernier. For example, you can read 1-minute of angle in either the vertical or horizontal planes on a 1-minute transit. You can also classify a transit by its magnification power, minimum focus distance, field of view, or stadia ratio.

The telescope on a transit is suspended between its standards. It rotates 360 degrees. A leveling vial is attached to the underside of the telescope. The cross hairs you see through the eyepiece of the telescope have one vertical hair and one

Courtesy: David White Instruments

Figure 4-4 Standard precise transit

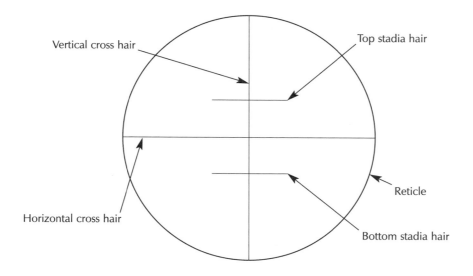

Figure 4-5 *Transit cross hairs*

horizontal hair at the center. Some have two additional short horizontal stadia hairs. The hairs are made of fine platinum wire. In the past, they were actually made of spider web! Figure 4-5 shows a typical arrangement of cross hairs in a transit or level.

You can use the stadia cross hairs to find the approximate distance from the instrument to a stadia rod. When the stadia cross hairs are preset to 1:100 and you read a 1-foot interval on the stadia rod, it means that the rod is 100 feet from the instrument.

A builder's transit is more rugged but less precise than a surveyor's transit. A builder's transit usually has a magnification power of 10, which means an object 120 feet away looks like it's only 12 feet away. A surveyor's transit may have twice that magnifying power.

The horizontal circle on a transit has two movable circular plates that rotate or lock. The two concentric plates rotate independently around the vertical axis of the instrument. The lower plate is a graduated circle. The upper plate has two verniers, set 180 degrees apart, and carries the standards and the compass box. The compass box is a rainproof enclosure that houses a magnetic compass. It's glass covered and sits in the center of the upper plate between the telescope supports. Use the two leveling vials, also called *spirit levels* or *plate levels*, mounted at right angles to each other on top of the horizontal circle to set the instrument on a level plane. If each bubble is in the center of the tube at the same time, the plates are horizontal.

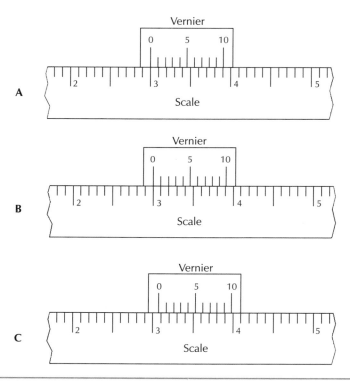

Figure 4-6 *Vernier*

The vertical circle contains two movable plates. One is attached to the side of a standard and the other to the telescope. You can read the angles between the plates on a graduated vernier. A vernier is a graduated scale used to measure parts of a circular scale. Most transits are graduated to read single minutes, but there are instruments that use verniers to read 30, 20, 10, and even 5 seconds.

The principle of the vernier is that the width of the divisions on the vernier is slightly less than the divisions on the scale. Figure 4-6 shows this principle. Note that the distance between 0 and 10 on the vernier is the same as the distance between 3.00 and 3.90 on the scale. When 0 on the vernier coincides with 3 on the scale, the reading is 3.00, as shown on Figure 4-6A. When 1 on the vernier coincides with 3.10 on the scale, the reading is 3.11, as shown on Figure 4-6B, and when 8 on the vernier lines up with 3.80 on the scale, the reading is 3.08, as shown on Figure 4-6C. The leveling base has a leveling screw you can use to adjust the horizontal circle relative to the base and tripod.

Most transits come with a high visibility case and a carrying strap. The case holds a sunshade you put on the end of the telescope, a dust cap to protect the objective lens when you're not using it, and a rain cover. There's also a brass plumb bob to center the instrument over a selected point, a magnifying glass to help you read the vernier, adjusting pins for tightening and correcting the leveling vials, a screwdriver for adjusting other components of the instrument, and an instruction book.

Here's how to set up a transit:

- Mount the transit on a wood or metal tripod.

- Hang the plumb bob, or plummet, from underneath the transit to a set point on the ground.

- Position the transit by setting the tripod firmly on the ground with the telescope approximately at eye level and the plumb directly over the intended point.

- Adjust the leveling screws so the compass box is horizontal and the plate stays level no matter what direction you point the telescope.

You'll need to know about angles, bearings, and azimuths to use a transit. A bearing is the angle between a line and the north-south line. It's never more than 90 degrees. For example, a line of 30 degrees 20 minutes west of north, is written as N 30° 20'W. A line 45 degrees 30 minutes east of south is written as S 45° 30' E. A bearing used in a property description is based on a previously recorded bearing, such as a street centerline or a section line. You should be able to add and subtract bearings to find angular distances, and to interpret the verniers on the horizontal and vertical circles to read the angles.

As an example, Figure 4-7 shows a typical parcel description by metes and bounds. Figure 4-8 shows how the bearing of each line was developed. The direction of a line is defined by the angle between that line and the north-south line. The traverse shown in this figure was described in a clockwise direction. If the direction were counterclockwise, the bearings would be reversed. The first course (between corners A and B) N 40° 15'E would be S 40° 15'W.

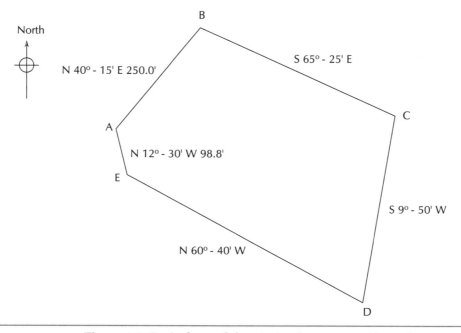

Figure 4-7 Typical parcel description by metes and bounds

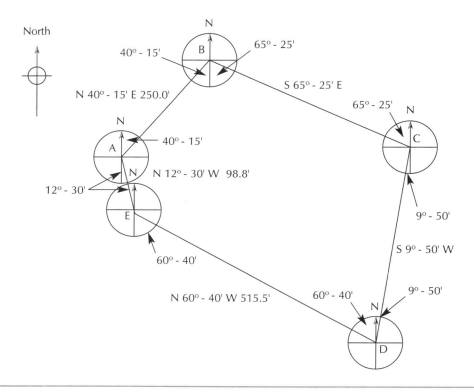

North

40° - 15'

65° - 25'

N 40° - 15' E 250.0'

S 65° - 25' E

65° - 25'

40° - 15'

N 12° - 30' W   98.8'

9° - 50'

12° - 30'

S 9° - 50' W

60° - 40'

9° - 50'

N 60° - 40' W 515.5'

60° - 40'

Figure 4-8 *Traverse by bearings and distances*

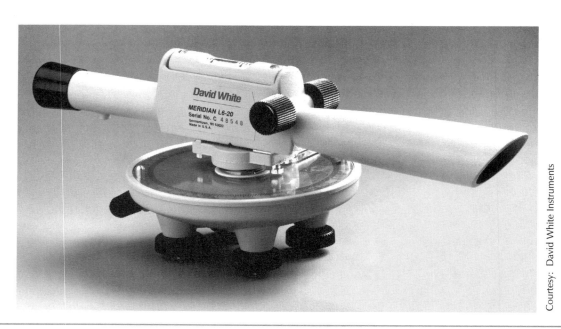

Figure 4-9 *Type of level used in construction*

Levels

A level is similar to a transit, except that its telescope is longer and has greater magnifying power. Figure 4-9 shows the type of level used in construction. A level is more practical than a transit in setting elevations because its telescope doesn't rotate vertically, it has fewer moving parts, and it's easier to keep steady. Most levels don't measure vertical or horizontal angles.

A basic level has a telescope with an attached level vial. The telescope is mounted on the ends of a straight bar. The bar rotates horizontally around its center on a vertical axis. Use the adjusting screws on the tripod to make the plate exactly horizontal. When you rotate the telescope on a level, you're sighting on a horizontal plane. Any object that is behind the horizontal cross hair in the eyepiece is at the same elevation as the telescope. Figure 4-10 shows different conditions when you do leveling work on sloping land.

A builder's transit-level serves as both a transit and a level. It's also more rugged than either a transit or level. The smallest angle you can read with a builder's transit-level may be 5 to 15 minutes.

The automatic laser level is an instrument that lets you do leveling work by yourself. See Figure 4-11. The level uses electronics and infrared technology to generate a reference plane around the instrument with a rotating emitter. It has a battery with a charger. Some laser levels have a hinged top so they can also generate a vertical reference plane. You use the bull's eye bubble to initially level the instrument. After that, it automatically keeps itself level. If you accidentally knock this instrument off level, it stops rotating and flashes a light.

Here's how to use an automatic level:

- Measure the height of the instrument with a yardstick you set between the ground point and a mark on the housing.

- Use the detector or sensor mounted on a staff to take measurements. The detector has a battery-powered display and tone generator.

- Read the sensor from front or back. Several people can use an automatic laser level at the same time.

Hand levels are handy to carry around but not accurate enough for precise work. You can use them for rough grading, but follow up with a leveling instrument mounted on a tripod.

You can make sure strings and batterboards are level using a portable leveling vial. You attach the vial to a tight string or to the top of batterboards to make sure they're horizontal. You can also use a 4-foot-long carpenter's spirit level to keep batterboards or forms level.

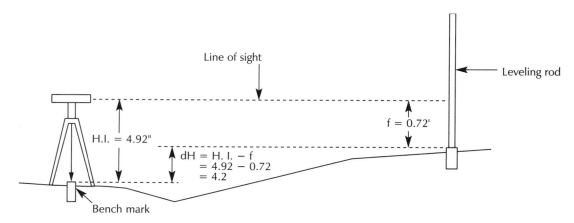

A Leveling with instrument over bench mark

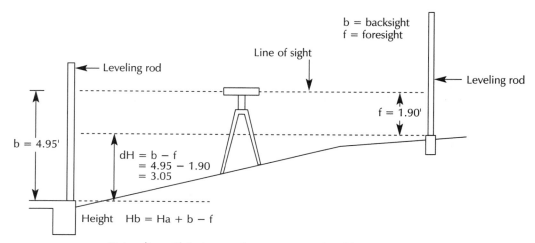

B Leveling with instrument between two rod positions

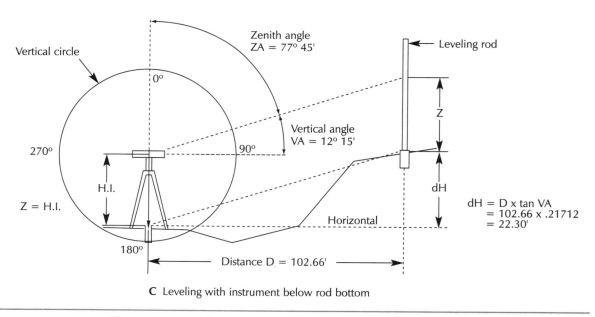

C Leveling with instrument below rod bottom

Figure 4-10 Leveling with a transit on sloping ground

Figure 4-11 Electronic automatic level

Rods

There are many types of leveling rods you can use to make vertical measurements. The most common type is the Philadelphia rod. It's made of hardwood to resist shrinking and swelling. The face is made of white Mylar with markings and numbers that won't rub off. Every number indicating a foot is in red, tenths of a foot are marked in black. Hundredths of a foot are marked in black bars along one edge of the rod. On a standard leveling rod, the numbers increase from bottom to top. On a direct elevation rod, the numbers increase from top to bottom.

A builder's rod is smaller but more ruggedly built than a Philadelphia rod. A builder's rod is usually 12 feet long. You can fold it into two sections. It's marked in feet (in white) and inches to the nearest eighth of an inch (in red). An engineer's rod is marked in feet and decimal parts of a foot.

Poles

The measuring pole is another type of rod. A typical model is an aluminum digital measuring pole that extends to 26 feet in height. It's self-reading, so you can measure both height and depth using only a measuring pole. You read distances on a digital dial attached to the lower section of the pole. Use this type of pole to measure the height of roofs and ceilings, or widths between walls. One person can use this instrument to make measurements in difficult locations where normally you'd need two people and a ladder.

Steel Tapes

Construction surveyors usually use steel tapes to measure distances. Tapes may be chain tapes, flat steel wire tapes, ordinary steel tapes, or metallic coated tapes. Most steel tapes are 100, 200, or 300 feet long. They're marked at 1-foot intervals. One end is marked in tenths and hundredths of a foot, reading from right to left. The 100-foot tapes are kept on reels when not in use. You can use leather thongs and holding clamps to keep a tape tight. Use tension handles to help reduce sag in a tape.

It's not a good idea to use cloth and metallic tapes for layout work. Cloth tapes can stretch too much, especially when they're wet. Metallic tapes also stretch more than steel tapes.

Measuring Errors

Most measuring errors are caused by:

- Reading a tape incorrectly. The most common reading error is omitting either 1 foot or 10 feet.

- Not holding a tape horizontal. Use a hand level to keep a tape horizontal.

- Not keeping a tape straight. To help keep a long measuring line straight, set temporary stakes in the ground along the measuring line.

- Putting the wrong tension on a tape. Normal tension on a 100-foot tape is 15 pounds. Most steel tapes are tested at the factory for the amount of tension required to accurately measure a 100-foot length. For example, the manufacturer may state that accuracy is within +/− 0.1 inch per 100 feet when the tape is supported throughout at 15-pound tension and 68 degrees F.

Figure 4-12 shows how to use a tape to measure distances on sloping ground.

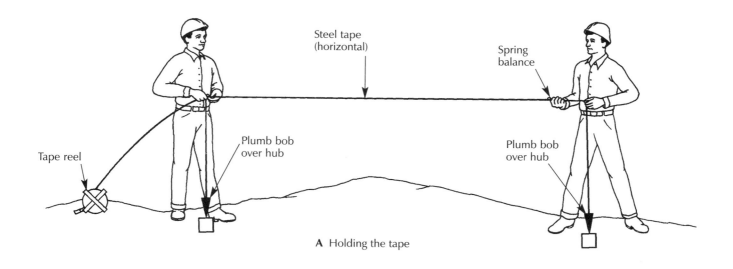

A Holding the tape

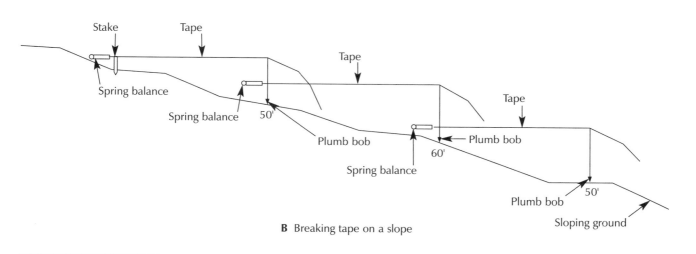

B Breaking tape on a slope

Figure 4-12 Using a steel tape on sloping ground

Foundation Layout

The first step in laying out a foundation is to surround the work with batterboards. Keep the batterboards clear of the foundation excavation. Set the tops of the batterboards at the same elevation as the top of the foundation wall. Drive a single nail into each batterboard at the extension of the outer lines of the foundation wall. When you connect these nails with a tight string, you have outlined the foundation (Figure 4-13).

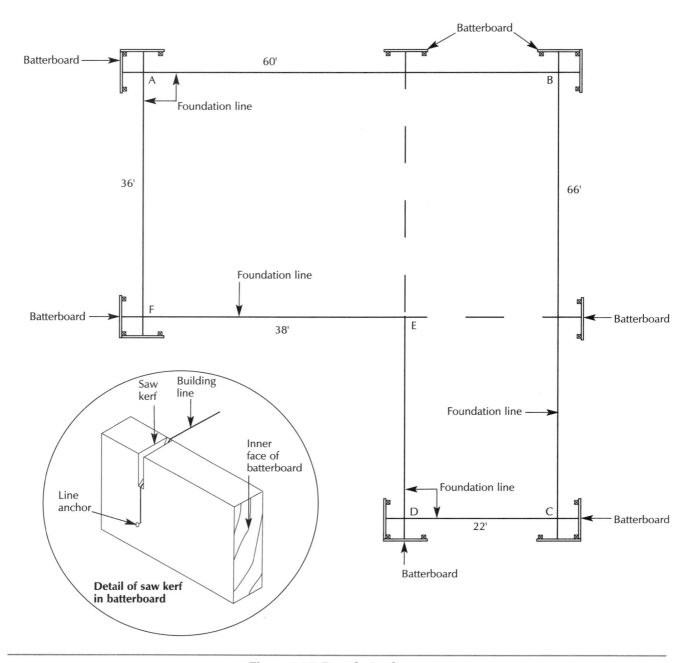

Figure 4-13 Foundation layout

You can check to see if a building layout is a true rectangle by measuring the diagonals between opposite corners of the layout. See Figure 4-14 for an example. The diagonals should have the same length. You can also check whether corners are 90 degrees by using a triangle with a 3-foot, 4-foot, and 5-foot side, or multiples like 6, 8 and 10 feet.

Use the elevation of the top of batterboards to measure down to the bottom of footings using a leveling rod and level. This establishes the depth of excavation.

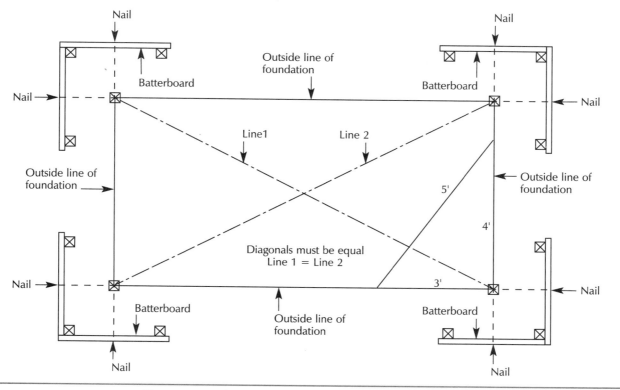

Figure 4-14 Checking diagonals and angles for squareness

Don't remove the nails in the batterboards. You'll use them to locate the wall forms after you're through with the footings. You usually use the outside surface of the foundation walls as a horizontal control of the building. Horizontal control means to be sure that the lengths of the perimeter or the outside foundation surfaces of a building are accurately measured and marked out.

Slab Openings

When you're about to place concrete for an elevated slab on forms and shoring, you should locate all sleeves, openings, and embedded items by surveying. Mark the plywood form with the centerlines of these items. Place the sleeves and forms for openings before placing the reinforcement and concrete. Use the outside surface of the perimeter walls as the base of the survey.

In summary, the construction surveying you do on a job is extremely important to help you lay out the job properly. In addition to telling you that you're laying the foundation in the right place, it's your last chance to check the plans for errors. A mistake that gets by will hurt just about everyone involved. Catching that mistake can save everybody's profits, and will earn you the reputation of being a thorough, conscientious professional.

Surveying might not be your favorite part of the job, but it's one of the most important parts.

Chapter 5

Concrete Formwork

Just because concrete formwork is temporary doesn't mean it can be done carelessly. You've got to design, engineer and build it accurately. The shape, position and finish of the concrete you pour will depend on the formwork you use. Formwork also has to withstand high pressures and loads without collapsing. Each part of a forming unit has to work together to make the entire system stable, watertight and resistant to the pressure of fresh concrete. You also need to build formwork efficiently and economically, without sacrificing quality or safety.

Form Materials

There are many materials you can use to build forms, including wood, plywood, steel, plastic, fiberboard, and certain types of composite wood board. The sheathing you use in a form is the mold for the outside surfaces of the concrete you pour, so make sure it's smooth, tight, and strong. And be certain that any wood you use is kiln-dried. Extremely dry wood that has a moisture content less than 19 percent may swell when it comes in contact with wet concrete.

Plywood

If you use plywood for sheathing, choose exterior grade $1/2$- to 1-inch plywood form sheathing with sanded surfaces. Be sure that any reconstituted wood core board you select can be used with fresh concrete. Always install plywood with the face grain perpendicular to the supporting members.

One popular type of sheathing is plywood overlaid with a plastic coating. This material is called high-density overlaid (HDO) plywood. It's made with a phenolic resin-impregnated fiber material that's applied to one or both sides of an exterior-type plywood core. HDO plywood is expensive, but it makes a smooth concrete surface and you can use it many times. Other types of plastic-clad plywood include medium-density overlay and Plyron.

Fiber-Reinforced Plastic (FRP)

You can also use fiber-reinforced plastic (FRP) for forms. It has a smooth surface and it's easy to strip. Plastic form liners provide a smooth or textured concrete surface. You can get liners that have vertically ribbed or fluted surfaces that give a board or brick appearance to concrete. Other types of FRP forms can replace board forms for one- and two-story buildings.

You can get custom designed column forms such as hexagonal-shaped ones that are made of FRP. Or you can use molded FRP one-piece column forms. You can get FRP column forms in sizes from 12 to 48 inches in diameter, and up to 20 feet long. These forms don't dent, sag, rot, or weather. They need little, if any, maintenance. They're lightweight, have only one vertical seam, and come with bracing collars. You can use specially designed bolts and nuts for fast and easy assembly and disassembly, or *stripping*, of the forms. Installation is simple: set up the form, pour the concrete, let it cure, strip the form, and you're done. And you can use these forms repeatedly.

Tubular Fiber Forms

You can also cast round columns using tubular fiber forms made of asphalt impregnated cardboard. Use these one-time fiber forms to cut the cost of pouring round columns, piers, piles, underpinning, and encasing wood or steel piling. You can quickly set, brace, and anchor these forms. Also, where beams tie in, you can easily cut and notch the forms to accommodate the beams. Here are some suggestions for handling, installing, and using round fiber forms:

- Store forms vertically and seal ends to keep out rain and snow.

- To place a form, lower it carefully over a reinforcing bar cage without damaging the inside surface of the form. A reinforcing bar cage is an arrangement of horizontal and vertical bars wired together as a single unit.

- Brace form with light timber or scaffolding. Use collars or wooden frames at the base of the form.

- Pour concrete continuously, up to 40 feet in height.

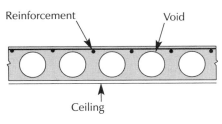

Reinforcement   Void

Ceiling

A Concrete floor slab with voids

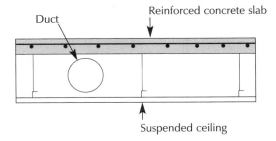

Duct   Reinforced concrete slab

Suspended ceiling

B Conventional floor slab with suspended ceiling

Figure 5-1 Concrete floor slabs

■ Strip form as soon as concrete has set. Use a power saw to cut two vertical slits, and remove form. Slit the form with a sharp knife to a depth of 12 inches from the top. Use a broad-bladed tool to pry the form off with short strokes by pushing the handle toward the column (or whatever item you're forming). Don't mar the concrete surface. After stripping, replace the form halves on the column and wire in place for further curing. This also protects the column during construction.

You can also use round fiber forms to create voids in floor slabs. Laying voided flat slabs will let you use shallower floors and give you a smooth underside for plastering, painting, or laying tile. Finally, you can use voids for heating and cooling system ducts. See Figure 5-1.

Fiberglass Forms

Use molded fiberglass forms for concrete waffle slabs, flat slabs, beams, and pan joist construction. Waffle slab forms are molded from metal dies and come in sizes from 19 to 40 inches square. They range in depth from 4 to 16 inches. They're easy to handle and install at a job site. You just set them on a clean plywood form, at the required spacing, to form the two-way joists. Joist widths vary from 5 to 8 inches. Also, you can remove and stack these forms easily. A blast of air will dislodge them from cured concrete.

Using Liners to Add Texture

When you want a patterned architectural concrete finish, use a sheathing with an inside finish that has the texture you want. You can do this by attaching a plastic liner to the inside of forms. If specifications call for aggregate finish, bond pea gravel to the sheathing with an adhesive such as epoxy.

Form liners are ideal for texturing tilt-up, cast in place, and precast architectural concrete. You attach the rigid plastic liners to the casting bed or formwork before you place any concrete. The smooth surface of the liner won't absorb moisture from the mix, or cause discoloration or voids. And, you can easily strip the liners from concrete. Liners come in different grades, ranging from a single use on up to 25 uses. You can use them on buildings, parking garages, bridges, retaining walls, planters, spandrels and feature strips. A typical sheet is 4 by 10 feet.

Composite Wood Panels

Composite wood sheathing panels include hardboard, waferboard, strand board, particleboard, and fiberboard. Most of these boards require 1 × 4 wood backing supported by studs. It's important to select composite wood panels that are specially manufactured to use for formwork. The edges must be sealed and the surfaces durable when exposed to wet concrete. You can use some composite wood panels for loadbearing formwork sheathing if you back them with 1 × 4s or 2 × 4s spaced 16 inches on center. Use weaker panels as liners to make smooth surface concrete.

Hardboard is a fibrous-felted board made of wood fibers impregnated with drying oils and bonded together with heat and pressure. You can use tempered hardboard impregnated with a drying oil as a form liner or facing material. Back up hardboard wall liners with 1 × 4s spaced 6 inches on center. Back up hardboard slab forms with 2 × 4s spaced 6 inches on center.

Waferboard is made with large thin wood wafers mixed with waterproof phenolic resin glue and bonded into thick mats with heat and pressure. You can use waferboards in place of plywood for wall and roof sheathing.

Strand board, also known as Oriented Strand Board, or OSB panels, are made of compressed wood particles arranged in three to five layers at right angles to one another. They usually have a polymer overlay that you can use in formwork, but always check the manufacturer's specifications to make sure.

Particleboard and fiberboard are made of wood particles and fibers bonded with synthetic resins. Fiberboard is made of wood or other plant fiber compressed into large sheets. Particleboard is made of a combination of wood particles and wood fibers bonded together with synthetic resins or other bonding material.

Wood Boards

You can build forms with wood boards or plywood at a job site. Use wood boards to form the edges of slabs and to construct complex forms for beams and girders. Use boards 1 or 2 inches thick, depending on the spacing between supporting studs and the pressure of the wet concrete. Wood board sheathing isn't used often now because it takes too long to install and it leaves an unattractive concrete surface finish after the forms are stripped. For a smooth surface, use T&G boards.

Form Studs

Use form studs to support sheathing. Form studs are usually 2 × 4s or 2 × 6s spaced 12, 16, 24, or 32 inches on center. The usual spacing is 24 inches. Maximum stud spacing depends on the type and thickness of the form sheathing, stud size, maximum height of pour, and ambient temperature. If three of these items are set, you can adjust the remaining item.

Here's an example. Let's say you have a maximum concrete pressure of 600 psf and a 2-foot stud spacing. The pressure on each stud is 600 × 2 or 1,200 plf. The span of the studs is determined by the spacing of the wales. Figure 5-2 shows the maximum span for lumber framing (the studs in this case). This table shows that the maximum span of 2 × 4 studs with 1,200 plf is 20 inches. If you use 2 × 6 studs, the maximum span is 29 inches. Check the American Plywood Association Publication V345 for valuable information on concrete form design and engineering data.

You can prefabricate form panels by nailing Plyform to the studs and the top and bottom plates. Prefabricated panels are normally 2 × 4, 2 × 8, or 4 × 8 feet. Prefabricated panels save time since you only need to install the wales, braces and ties to complete the formwork.

Uniform load (plf)	Continuous over 2 or 3 supports (1 or 2 spans)							
	2 x 4	2 x 6	2 x 8	2 x 10	2 x 12	4 x 4	4 x 6	4 x 8
200	49	72	91	111	129	68	101	123
400	35	51	64	79	91	53	78	102
600	28	41	53	64	74	43	63	84
800	25	36	45	56	64	38	55	72
1000	22	32	41	50	58	34	49	65
1200	20	29	37	45	53	31	45	59
1400	19	27	34	42	49	28	41	55
1600	17	25	32	39	46	27	39	51
1800	16	24	30	37	43	25	37	48
2000	16	23	29	35	41	24	35	46

Spans based on single member allowable stress multiplied by a 1.25 duration of load factor for 7-day load. Deflection limited to $1/360$ of the span with $1/4$" maximum. Spans are center-to-center of supports. Douglas fir No. 2 or Southern pine No. 2.

Figure 5-2 Maximum spans for lumber framing (inches)

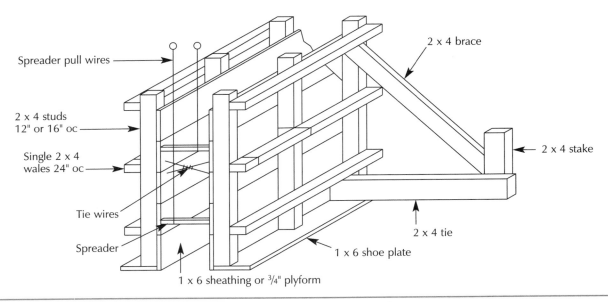

Figure 5-3 *Formwork for a wall*

Shoe Plates

Use shoe plates (or sole plates) to level the bottom of forms and to anchor the bottoms of studs in. Toenail or end-nail the studs to the shoe plate. See Figure 5-3. A shoe plate is usually a 1×6, 2×4, or 2×6.

Form Braces

Bracing helps keep forms aligned and rigid. A typical brace is a 2×4 you install diagonally from a wale or stud near the top of a form to a stake driven into the ground. For added stability, install a horizontal member from the stake to a stud or shoe plate. See Figure 5-3.

Wales

You can use studs and braces to support walls less then 4 feet high as shown in Figure 5-3. But you need to use wales (walers) to support the studs for walls over 4 feet high as shown in Figure 5-4. Make wales of single or double 2×4 lumber for light loads, and 2×6s for heavy loads. The wale spacing you use will depend on the strength of the studs and the lateral pressure of the concrete.

Stakes

Stakes can be 2×4s driven into the ground, or you can use pointed steel stakes. Use steel stakes for curbs, gutters, footings, foundations, and similar work. These stakes have holes drilled 2 inches apart at 90-degree angles for nailing. Use

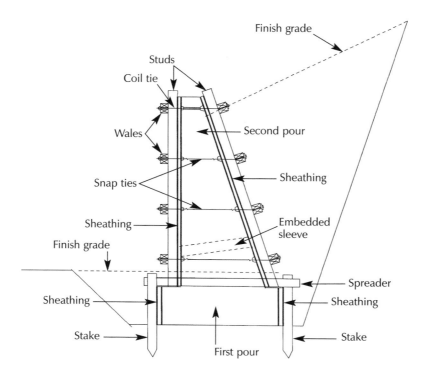

Finish grade

Studs

Coil tie

Wales

Second pour

Snap ties

Sheathing

Sheathing

Embedded sleeve

Finish grade

Spreader

Sheathing

Sheathing

Stake

Stake

First pour

A Gravity retaining wall

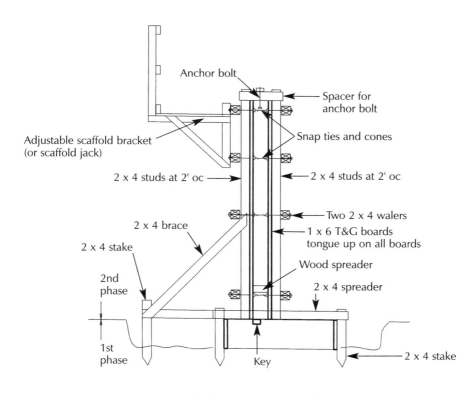

Anchor bolt

Spacer for anchor bolt

Adjustable scaffold bracket (or scaffold jack)

Snap ties and cones

2 x 4 studs at 2' oc

2 x 4 studs at 2' oc

2 x 4 brace

Two 2 x 4 walers

2 x 4 stake

1 x 6 T&G boards tongue up on all boards

2nd phase

Wood spreader

2 x 4 spreader

1st phase

2 x 4 stake

Key

B Concrete retaining wall

Figure 5-4 Forms for retaining walls

flat curb stakes to set curb forms and foundations. Round steel stakes have drilled holes at 1-inch spacing in a spiral pattern 60 degrees apart. All these metal stakes come in lengths of 18, 24, 30, and 36 inches. It's a good idea to use a stake puller to remove stakes without damaging them.

Spreaders

Spreaders are removable pieces of wood you set between sheathing to hold the sheathing apart. Spreaders are held in place by friction, but you can pull them out through fresh concrete by an attached wire. See Figure 5-3. If you use snap ties, you don't need spreaders.

Form Ties

The main purpose of form ties is to keep the sides of a wall form from spreading apart. Many builders just use twisted wire ties on simple structures where the strength of the tie and appearance isn't important. Where appearance is important, don't leave any exposed metal on the exterior surface. Some job specifications require that all embedded metal be at least $1^1/2$ inches from the concrete surface. You can also use nonmetallic ties, such as fiberglass.

Attach ties to each stud at the wales on both sides of forms. Tighten ties by twisting them with a wedge. Figure 5-5 shows form ties.

Form tie wire is usually made of 8- or 9-gauge soft black annealed iron wire. If you find that the wire isn't strong enough to carry a load, leave studs in their original positions and space wales closer together. The approximate breaking strength of double-strand steel wires is:

8 gauge	1,700 lbs.
9 gauge	1,420 lbs.
10 gauge	1,170 lbs.
11 gauge	930 lbs.

Here's the formula you can use to figure out maximum tie spacing:

$$maximum\ tie\ spacing\ (in.) = \frac{wire\ strength \times 12}{uniform\ load\ on\ wales}$$

For example, if the maximum pressure is 600 psf and wales are 2 feet on center, the uniform load on each wale is $600 \times 2 = 1,200$ plf. Since 9-gauge wires have a breaking strength of 1,420 pounds, the maximum spacing of the ties is:

$$maximum\ tie\ spacing\ (in.) = \frac{1,420 \times 12}{1,200} = 14.2\ inches$$

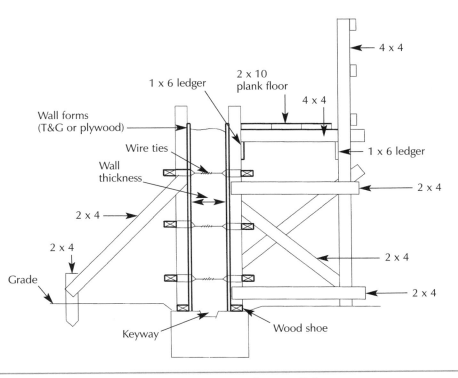

Figure 5-5 *Runway along wall form*

To install twisted wire ties:

■ Bore small holes through the sheathing at each side of the stud where the wire tie will go.

■ Run the wires through both sides of the forms and around the wales.

■ Pull the wires tight and twist the ends together.

■ Place wood spreaders between the forms and twist the tie wires with a stick until the form walls are tight against the spreaders.

Snap Ties

Snap tie is a general term that describes any tie that you can break off at a specified depth. After you snap a tie by twisting the end with a wrench, you should patch the tie end with a waterproof nonshrink material that bonds well with concrete. If moisture reaches a metal tie end, you'll soon get rust stains on the concrete surface. Use nonshrink patching grout or dry pack mortar.

Some types of snap ties are:

■ snap ties with small cone spreader

■ washer spreader, crimped with break back

- cone spreader
- taper tie
- strap tie
- loop end tie used with panels
- straight tie with attached plastic tube
- threaded bar with unattached plastic sleeve
- coil tie with or without cone spreader
- crimped tie with disconnecting ends
- plain tie with shebolt disconnecting ends

You can break most snap ties within the concrete surface after you strip the forms. Figure 5-6 shows how to break or snap different ties free from the forms. In section A, you use a wrench head to remove them. In section B, you snap the ties at the weakened point, which may be a crimp or purposely-flattened portion of the tie that will break or snap when you twist the end of the tie. This releases the form from the cured concrete. In section C, you snap the tie at the crimp. In section D, you unscrew the tie ends from the shebolt that's embedded in the concrete. The shebolt, indicated in the figure, allows you to unscrew the end of the tie from the embedded portion.

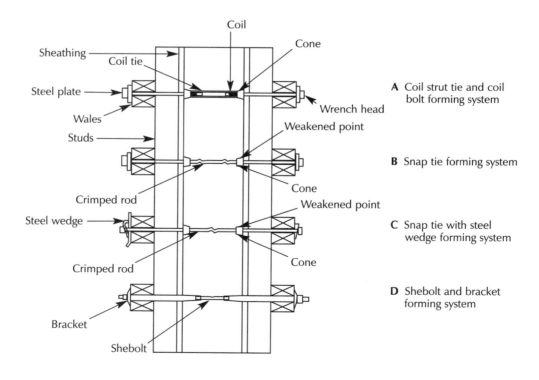

Figure 5-6 *Types of form ties*

Here are some safety rules for using snap ties:

■ Don't use bent snap ties.

■ Don't climb on installed snap ties.

■ Don't let snap ties stay in concrete more than 24 hours after placement. Remove at the break point as soon as possible.

■ Don't weld ties to any object.

■ Don't exceed safe work loads.

Form Hardware

Form hardware includes nails, screws, bolts, inserts, and sleeves. Use common nails to assemble form panels and other components that won't be disassembled when you strip the forms. Use 6d nails for 1-inch sheathing or $5/8$-inch plywood, and 3d blue shingle nails for fiberboard and thin plywood liners. Use double-headed nails to install kickers, blocks, braces, wales, or any other item you have to remove when you strip forms. Use bolts and lag screws for heavy formwork requiring 2-inch sheathing supported with heavy timbers.

Inserts

When working with concrete, you'll often have to use metal inserts to fasten beams and supported items such as pipe and conduit clamps, lifting lugs, and anchors. Some inserts accommodate threaded bolts, while others have a slot you insert a bolt head into. Sometimes you must use embedded weld plates so you can weld a steel member to the concrete. Most inserts are the female part of a bolted connection that's embedded in the concrete. To prevent concrete from entering an insert, plug the insert with a shaped plastic foam. Later, you can dissolve the foam with a petroleum-base solvent.

The most common insert is the coil pick-up insert. This is a helix coil with connecting wires or rods. The helix coil acts as a nut that you thread coil bolts into. The threads on these bolts are shaped to fit the wire's contour, which form the coil. The coil bolts are cone-pointed, and can be easily threaded into the coil. These inserts come in a wide range of sizes and configurations. They're made for $3/4$-inch to $1^1/2$-inch bolts.

When you construct tilt-up buildings, you'll need special inserts. Tilt-up construction is a technique where you precast concrete wall panels on a building floor slab or other casting surface. Then, you "tilt-up" the panels with a mobile crane and carry them to their final position. To do this, you'll need inserts that you can quickly and safely set bolts into, such as:

■ pick-up inserts that link a crane's lifting cables to the precast panels.

■ stiffening or "strongbacks" to help keep a panel from sagging as it's lifted from the floor slab or casting platform.

■ face inserts that connect eyebolts used for lifting a panel off the floor, for bolting the temporary braces that hold a panel upright, or for bolting wood ledgers to a concrete panel.

Sleeves

Set sleeves in forms to hold piping, conduit, and ductwork. Then you won't have to core through hardened concrete and run the risk of cutting steel reinforcing bars. Make sleeves from galvanized steel sheets, plastic, or cardboard. Use wood to form square or rectangular sleeve openings.

Concrete Pressure

When you place fresh concrete in a form to make a wall, it's in a semi-liquid state that presses out against the form in all directions. Your job is to design and build forms that can resist this pressure. The pressure at any location on a wall depends on several factors, but surprisingly, the thickness of the wall isn't one of them. These are the factors that determine the pressure at any location on a wall:

■ Weight of the concrete. Regular liquid concrete weighs between 150 and 160 pounds per cubic foot. Lightweight concrete weighs between 90 and 110 pcf, depending on the type of coarse aggregate in it.

■ Height of the liquid concrete. You can find the pressure at any point on the forms by multiplying the weight of the concrete by the distance from the point to the top of the liquid concrete.

The maximum pressure is at the bottom of a form and decreases toward the top of the form. Look at the hypotenuse (inclined leg) of the right triangles in Figure 5-7 beside each pour, or lift in the figure. Figure 5-7A shows the first 4-foot-high pour of three pours. Section B shows the second 4-foot-high pour, and section C shows the third 4-foot pour. Section D shows the concrete pressures for the entire 12-foot height. Sections E, F, and G are similar except that there are just two pours. Each pour is 6 feet high. The total height of wall is 12 feet. Section H shows the difference in pressure between regular and lightweight concrete.

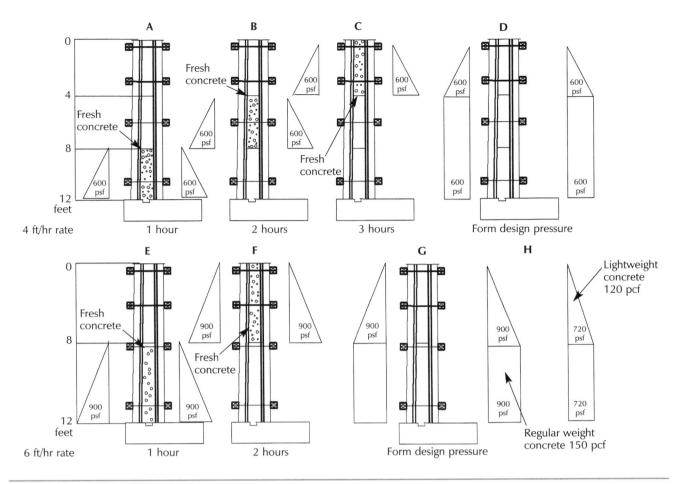

Figure 5-7 Form design pressures

■ Rate of pour. A fast rate of pour makes a form support all the liquid concrete, whereas a slow pour rate lets the concrete at the bottom of the form set up and become self-supporting before all the liquid concrete reaches the top of the form. Figure 5-8 shows how the concrete pressure on a form increases as the rate of pour is increased. For example, when you pour concrete at 70 degrees F and a pour rate of 5 feet per hour, it puts a maximum pressure of about 840 psf on the lower portion of the form. If you pour concrete at 70 degrees F and a rate of 7 feet per hour, it puts about 1,100 psf on the form. Figure 5-9 shows maximum concrete pressure and rate of placing concrete.

■ Surrounding temperature. Concrete sets up faster at higher temperatures. As concrete sets, pressure is reduced since more of the concrete is self-supporting. Figure 5-8 shows how concrete placed at the rate of 6 feet per hour exerts a maximum pressure of 980 psf at the lower portion of the pour at 70 degrees F, and approximately 1960 psf at 30 degrees F.

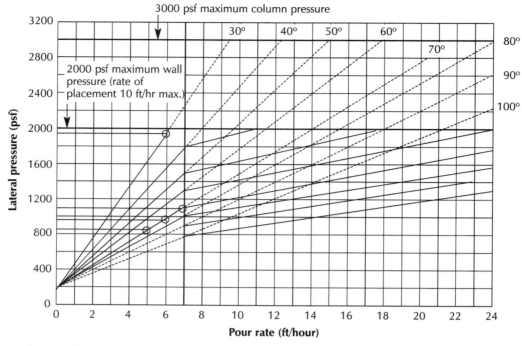

Figure 5-8 *Lateral concrete pressure at various temperatures*

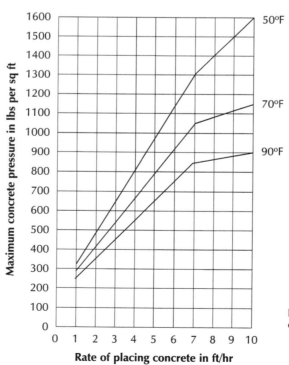

Note: Rate of placing ft/hr = Mixer output cfh/plane surface sq ft

Figure 5-9 *Maximum concrete pressure*

- Intensity of vibration of the mix. Vibration helps consolidate concrete, but it also slows the setup process, so it increases pressure.

- Concrete slump. A higher slump mix increases pressure because it's heavier and sets up more slowly than a lower slump mix.

- Types of chemical additives in the concrete. Some chemical additives, such as calcium chloride, speed up the setup process, so they decrease pressure.

Wall Form Design

Concrete details in plans and specifications should show the finished shape of the concrete and the texture of the exterior concrete surface. On major jobs, engineers have to design and approve forms and shoring, but on most jobs, carpenters design and build the formwork.

An Example of Wall Form Design

Let's go through a comprehensive example for the design of a typical wall form. The wall is $8^1/_2$ feet high and 10 inches thick. The concrete pour rate is 4 feet per hour. According to Figure 5-9, the maximum concrete pressure is 600 pounds per square foot at 80 degrees F.

Let's say that the studs will be spaced 1 foot apart to support plywood sheathing every 12 inches. According to Figure 5-10, Plyform Class 1 panels with a 600 psf load can be either $^{23}/_{32}$ or $^3/_4$ inch thick. The face grain of the plywood should be perpendicular to the studs. We'll use $^{23}/_{32}$-inch thick Plyform Class 1 panels.

Support spacing (in)	Plywood thickness					
	$^{15}/_{32}$	$^1/_2$	$^{19}/_{32}$	$^5/_8$	$^{23}/_{32}$	$^3/_4$
4	2715	2945	3110	3270	4010	4110
8	885	970	1195	1260	1540	1580
12	335	405	540	575	695	730
16	150	175	245	265	345	370
20	—	100	145	160	210	225
24	—	—	—	—	110	120
Face grain across supports, deflection limited to $^1/_{360}$ of the span and plywood continuous across two or more supports.						

Figure 5-10 Recommended maximum pressures on Plyform Class I (psf)

Each stud must resist a uniform horizontal load of 600 plf, since they're spaced 1 foot apart. The studs are supported laterally by wales. Let's see if we can space the wales about 48 inches apart. Looking back to Figure 5-2, 2 × 6 studs resisting a load of 600 plf shouldn't span over 41 inches. So we'll use 2 × 6 studs at 12 inches on center supported by wales spaced 41 inches on center. Make the wales out of doubled 2 × 6 lumber. Here's how to calculate the concrete pressure on the wales:

$$\text{pressure} = \frac{600 \text{ psf} \times 41}{12 \text{ ft}} = 2,050 \text{ plf}$$

Since we'll use double wales, each wale carries 2,050 / 2 or 1,025 plf. Form tie spacing controls the wale span. Figure 5-2 shows that 2 × 6 lumber spanning 32 inches can support a load of 1,000 plf, or spanning 29 inches can support 1,200 plf. Since our 1,025-plf load is in between those figures, we'll set the form ties at 30 inches on center. We use 30 instead of 29 inches because it's common practice to use increments of 2 or 4 inches spacing wales and ties.

Now we have to select the form ties. The load on each tie is the load on the double wale times the spacing in feet:

$$\frac{2,050 \text{ plf} \times 30 \text{ in}}{12 \text{ ft}} = 5,125 \text{ lbs}$$

Wire ties won't work because an 8-gauge wire is only good for 1,700 pounds. We must either reduce the spacing of the wire ties or select a manufactured form tie. According to their manufacturers, here are the capacities of some snap ties:

- a $9/16$-inch shebolt has a working load of 6,500 lbs.

- a $3/4$-inch taper tie is good for 10,000 lbs.

- a $1/2$-inch coil tie has a safe work load of 6,750 lbs.

Let's use $9/16$-inch bolt ties at 30 inches on center.

Rules and Checklist for Designing Forms

You should brace wall forms, in either direction, against wind. Check with your local building department for wind loads in your area. A common wind load is 20 psf. Inclined bracing on one side of the forms is usually adequate for wind in either direction. To resist wind uplift on a wall form, make sure you nail the studs to the sole plate and anchor the sole plate to the concrete footing. Attach the sheathing to the studs with as few nails as possible to make it easier to disassemble the forms. You may need extra nails on gang forms. Use 6d nails for $23/32$-and $3/4$-inch plywood. Don't butt the panels too tightly together as plywood has a tendency to swell the first time it's exposed to wet concrete.

Here are some rules you should follow when designing wood forms:

■ Determine what form materials are readily available.

■ Use stock sizes and lengths of lumber.

■ Use as few lengths of lumber as possible.

■ Use as few units as possible, but don't make the units too heavy.

■ Design the forms for easy stripping.

■ Try to use such units as wall panels, floor panels, and beam and column forms so that you can reuse them.

■ Make bevel cuts and keys so you can release forms with little prying.

■ Find out the rate of delivery of concrete to the job site in cubic feet.

■ Calculate the square feet of floor area the concrete will cover.

■ Figure the rate of pour in the forms in vertical feet per hour by dividing the rate of delivery (cf/hr) by the wall cross-sectional area (sq ft).

■ Try to forecast the air temperature at the time you place the concrete. During cold spells, you may have to use oil-fired heaters to heat the forms, and enclose the work with tarps.

■ Use Figure 5-9 to find the maximum concrete pressure in the forms.

■ Figure the maximum stud spacing.

■ Calculate the uniform load on each stud in plf.

■ Determine the maximum wale spacing. Wale spacing depends on the strength of the studs and the lateral pressure of the concrete.

■ Figure the uniform load on the wales in plf.

■ Find out the maximum tie spacing and tie strength required.

■ Compare the maximum tie spacing with the maximum stud spacing. The strength of tie wires is usually adequate for any spacing of the studs or wales.

Common Defects in Concrete Forming

Some common causes of defects or failure in concrete forming are:

■ inadequate diagonal and lateral shore bracing

■ placing concrete too quickly

■ unstable or frozen soil beneath mudsills supporting the shores

- insufficient nailing

- shoring not plumb

- unsecured locking devices on metal shoring

- vibration from concrete vibrators or moving loads such as concrete buggies

- supports or forms removed too soon

- knotholes, cracks, or other blemishes from the sheathing surface

- careless reshoring

- inadequate size and spacing of shores or reshores

- faulty formwork design

Recommended Tolerances When You Build Forms

Concrete forms will vary from plan dimensions due to errors in building forms, and the formwork settling and shifting. Some variations are acceptable. See the tolerances American Concrete Institute (ACI) recommends for formwork in Figure 5-11.

Safety Rules for Shoring

Here are some safety rules you should follow when you install shoring:

1. Follow local ordinances, codes, and regulations for shoring.

2. Use the manufacturer's recommended safe working loads for the type of shoring frame and height from the supporting sill to the formwork. To prevent accidents, the shore designer must consider the weight of wet concrete and forms, the strength of the stringers, joists, and sheathing, as well as the capacity of the shores.

3. Keep a shoring layout at the job site. A shoring layout is a drawing that shows the location and size of shores, bracing, stringers, and joists. It's usually prepared by the job superintendent or the job engineer. When you rent shoring equipment, the rental company usually prepares the shoring layout. Don't exceed the shore frame spacing or shore heights shown on the shoring layout.

4. Provide shoe plates to maintain a solid footing under each shore to distribute loads properly. Be sure that the soil under each shoe plate is stable.

5. Use adjustment screws to adjust the shore heights over uneven grade conditions.

6. Plumb and level all shoring and formwork before, during, and after you pour the concrete.

■ The variations in concrete plan dimensions should be within $-^1/_2$ and +2 inches (from $^1/_2$ inch less than, to 2 inches more than, the plan dimension).

■ The misplacement of concrete members should be within 2 percent of the footing width, but not more than 2 inches.

■ The reduction in thickness of members should be within 5 percent of the specified thickness.

■ The variation in plumb for concrete walls shouldn't be more than ± 1 inch for structures up to 100 feet high.

■ The variation in plumb for conspicuous lines such as control joints shouldn't be more than ± $^1/_2$ inch for walls up to 100 feet high.

■ The variation in any wall opening dimension should be within $-^1/_4$ and +1 inch.

■ The variation in wall thickness should be within:

   $-^1/_4$ and $+^3/_8$ inch for walls up to 12 inches thick
   $-^3/_8$ and $+^1/_2$ inch for walls 12 to 36 inches thick
   $-^3/_4$ and +1 inch for walls over 36 inches thick

■ The elevation of a concrete soffit should be within ± $^3/_4$ inch of the specified elevation.

■ The variation in an elevated slab's thickness should be within:

   $-^1/_4$ and $+^3/_8$ inch for slabs up to 12 inches thick
   $-^3/_8$ and $+^1/_2$ inch for slabs 12 to 36 inches thick
   $-^3/_4$ and +1 inch for slabs over 36 inches thick

■ The variation in any slab opening dimension should be within $-^1/_4$ and +1 inch.

■ The variation from level for a beam soffit should be within ±$^3/_4$ inch over the entire length.

■ The variation from level for an exposed parapet should be within ±$^1/_2$ inch over the entire length.

■ The deviation from any cross-section dimension should be within:

   $-^1/_4$ and $+^3/_8$ inch for thicknesses less than 12 inches
   $-^3/_8$ and $+^1/_2$ inch for thicknesses 12 to 36 inches
   $-^3/_4$ and + 1 inch for thicknesses over 36 inches

Figure 5-11 Recommended ACI tolerances for forms

7. Fasten all braces securely and constantly watch for potential form failure, especially when you pour the concrete.

8. Have the reshoring procedure approved by a qualified engineer.

9. Don't remove braces or back off adjustment screws until the job superintendent or engineer authorizes you to do so.

Wall Forms

Some types of wall forms are:

■ single wale with snap ties

■ double-wale

■ multi

■ conventional

■ modular

■ gang

■ crane set

Generally, you follow these steps to install a single wale form:

1. Drill holes in the plywood panels to put the ties in.

2. Install a 2 × 4 kicker plate on the top of the foundation.

3. Install drilled plywood panels by nailing bottom edges to kicker plates.

4. Install wrench head snap ties through predrilled holes in the plywood panels.

5. Install wale brackets.

6. Install wales.

7. Set opposite plywood panels and lock on snap ties.

To add height to a form, install another 2 × 4 kicker plate on top of each panel.

A *double wale unit* has wales on both sides of the forms. Otherwise, it's the same as a single wale wall form.

Ganged forms are prefabricated wood or metal panels complete with studs, wales, and ties. The average panel size is 8 × 6, but a prefabricated form can be as large as 12 feet high and 16 feet wide. Here's how to install ganged forms:

1. Lay out the walls.

2. Fasten the bottom plates.

3. Handle and set the wall panels.

4. Brace and align the wall panels.

5. Adjust the ties.

6. Install the steel reinforcing bars.

Crane set forms are prefabricated forms that are placed in position by mobile cranes.

You can use *floor forms* for casting elevated concrete slabs. They may be of plywood, steel, or fiber reinforced plastic (FRP). The steel and FRP forms are used for casting one- and two-way joist and waffle slabs. There are special forms for columns, slabs, decks, and precast concrete, such as tilt-up.

An *earth form* is the cheapest type of form. Generally, you can use one for footings and foundations if the soil is stable enough to retain its shape. One advantage of earth forms is that they require less excavation and they don't settle a lot. The main disadvantage is that they make a rough surface on concrete. That's why they're often used only for below-grade work.

Figure 5-12 shows typical foundations and footings for one- and two-story buildings. Use this figure only as a guide for footing types, since there are many things you need to consider about which size footing you should use. For example, you have to find out the depth of the frost line in the area you're building in. Foundations and footings must extend below the frost line. Check with your local building department regarding the frost line. In some parts of the northern United States, the frost line is 72 inches below grade.

A wall form may use sheathing, studs, wales (or walers), braces, stakes, shoe plates, spreaders, and form ties. Snap ties are used more often than form ties because they're easier to install. Figure 5-13 shows the typical parts of formwork for a grade beam. Figure 5-3 shows one way you can make a form for a concrete wall.

Usually you'll need a walkway on the side of a high wall form. To save time building a walkway support, use steel adjustable jack brackets. Use triangular-shaped brackets to support a wood walkway and posts for a guardrail.

You can attach the brackets to a cured concrete wall after you strip the walls by connecting the brackets to embedded inserts. Then you can use these brackets to support a walkway for access to fill any holes, voids or other blemishes in the concrete. Figure 5-4 shows an example of a retaining wall form that has wall surfaces that are battered or inclined.

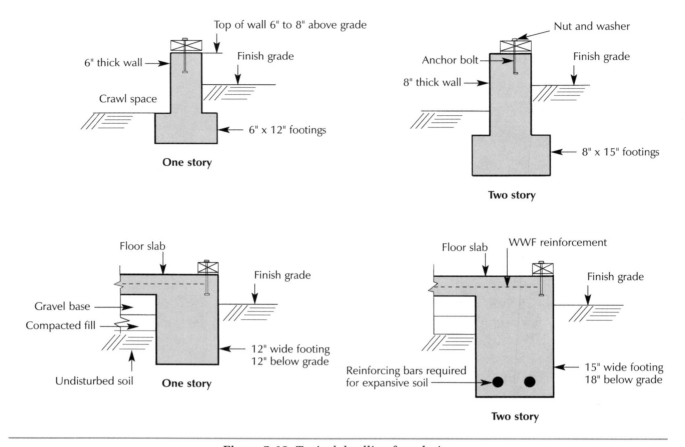

Figure 5-12 Typical dwelling foundation

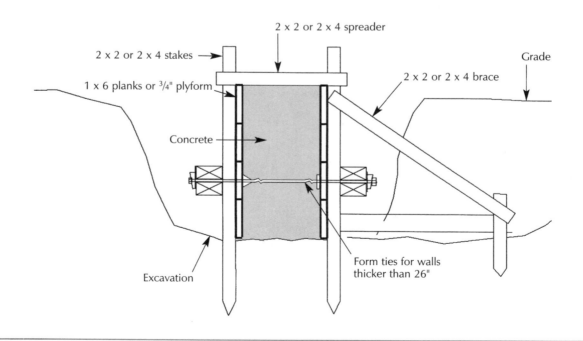

Figure 5-13 Formwork for grade beam

Column Forms

Figure 5-14 shows typical forms for a pile cap, pier and column. The major parts of column forms include:

- templates set on the foundation for positioning the column forms.
- forms made of wood, prefabricated steel, a combination of steel and plywood, or fiber reinforced plastic (FRP).
- bracing to stabilize and align the forms.

There are also some accessories you've got to use with column forms. Chamfer strips smooth out the sharp corners of rectangular columns. Use wood or steel yokes and locks to hold sheathing in place and resist the pressure of the concrete.

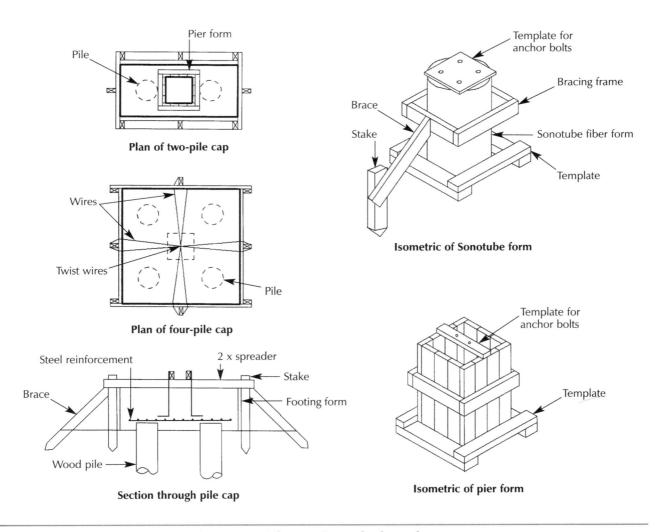

Figure 5-14 Pile cap, pier, and column forms

Largest dimension of column (inches)								
Height (ft)	16	18	20	24	28	30	32	36
1								
2	31	29	27		21	20	19	17
3				23				
4		28	26		20	19	18	15
5								12
6					18	18	17	11
7	30			22	15		13	10
8		26	24		13	12	12	
9	29			16	12	10	10	8
10		20	19					
11			16	14	10	9	8	7
12	21	18	15	13	9	8		
13	20	16	14	12			7	6
14					8	7		
15	18	15	12	10		6	6	
16	15	13	11		7			
17	14			9	6			
18		12	10	8				
19	13							
20	12	11	9					

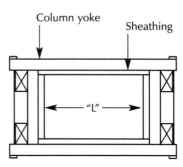

Column yoke · Sheathing · "L"

Figure 5-15 *Maximum spacing of rectangular column form yokes*

For a rectangular column, space yokes vertically according to the largest cross-sectional dimension of the column and its height. You can space yokes from 12 to 30 inches for columns 16 to 36 inches wide and up to 20 feet tall.

Figure 5-15 shows the suggested spacing of column yokes. Here's how to use the table to find the maximum spacing of column yokes:

- Select the height of the column along the left side of the table.
- Select the largest cross-sectional dimension of the rectangular column at the top of the table.
- Find the point where the column values intersect in the table.

For example, if a rectangular column is 12 feet tall and its largest cross-sectional dimension is 18 inches, the maximum yoke spacing is 18 inches.

You can use steel column clamps instead of wood yokes. Since steel yokes require the same spacing as wood yokes, you can use Figure 5-15 to find the spacing you need for various types of metal column clamps.

Here's how you design and build a column form:

1. Determine what form materials are readily available.
2. Determine the height and cross-sectional dimensions of the column.
3. Determine the yoke spacing.
4. Build and install a template at the base of the column.
5. Build and erect the column forms and yokes.
6. Brace and align the column forms in both directions.

Column Anchor Bolts

Install wood templates to hold column anchor bolts in position in slabs or concrete walls. Secure each bolt to the template with a pair of nuts, one below and one above the template. Carefully locate the template using a steel tape and chalk line. Then remove the upper nut and template after the concrete has set up, leaving the projected end of the threaded bolt above the top surface of the concrete. Figure 5-16 shows the tolerances allowed for column anchor bolts. You can install anchor bolts connected to hold-down hardware with a $1/8$-inch tolerance since the bolts are inserted into slightly oversized holes.

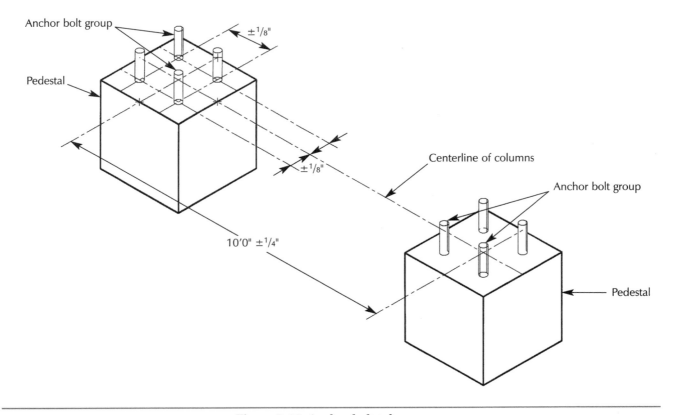

Figure 5-16 *Anchor bolt tolerance*

Elevated Slab Forms

The major components you need to form an elevated concrete slab include shores, beams, stringers, and sheathing. Supports and forms for concrete stairs and balconies are similar as they both involve posts, joists, and sheathing. See Figures 5-17 and 5-18.

You can use 4 × 4 wood posts or tubular steel shores to support the beams. You can adjust the height of a wood shore, or post, with a turnbuckle assembly or Ellis clamps.

Figure 5-19 shows metal shoring. Steel shores have telescoping steel tubes you can use to vary the height of a shore from 6 to 14 feet. You can turn a screw at the top of a shore to fine-tune the final height of the shore. The manufacturers of various tubular shores provide tables showing the capacity of each type of shore at various heights. The longer the shore, the less it can support. For example, a wood shore 6 feet tall can support 7,500 pounds, but a 14-foot shore can support only 2,000 pounds.

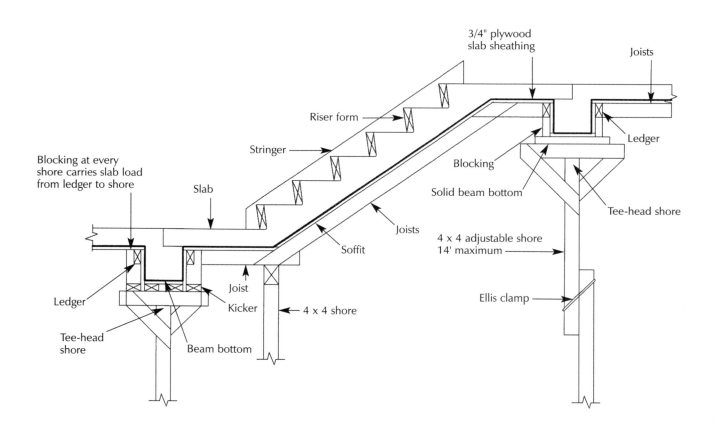

Figure 5-17 *Forms for concrete stairs*

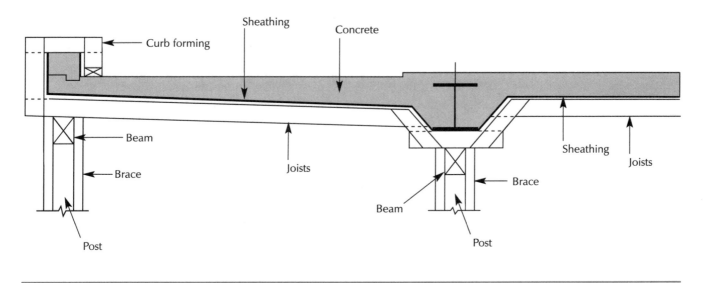

Figure 5-18 *Forming a concrete balcony*

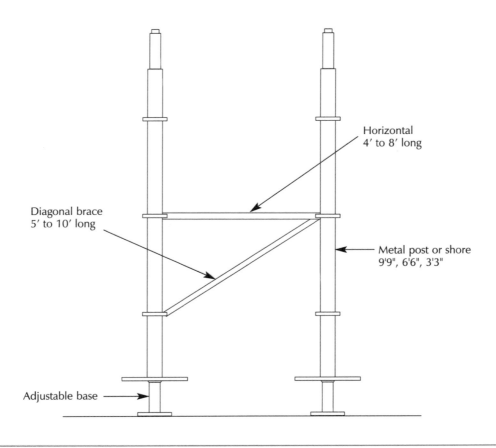

Figure 5-19 *Metal shoring*

Beams support stringers (joists) which, in turn, support sheathing. Beams can be solid wood, laminated wood, box beams, or I-beams. Some beams are made of lightweight aluminum that you can adjust from 4 to 20 feet long.

You can use 1 × 6 tongue-and-groove boards, or plywood sheathing, to form the soffits of elevated slabs. Or you can use ribbed or corrugated steel decking.

Here's how you erect forms for an elevated slab:

1. Erect 4 × 4 posts or steel shores.

2. Install mudsills if the shoring is supported on earth.

3. Install beams over the shores and bracing between the shores. Brace steel shores with steel rods, angles, or channels.

4. Install stringers perpendicular to the beams.

5. Place the sheathing over the stringers.

Figure 5-20 shows the design loads for slab forms. These values include a 50-psf load for workers, nonmotorized equipment, and impact. They also include the weights of beam forms, stringers, and sheathing. If you use motorized buggies, add 25 psf to the load.

Reshoring an Elevated Slab

Reshoring is when you reinstall supports for an elevated slab form after you've partially stripped the form. In some cases, where you've stripped the forms to reuse them, you'll have to reshore the concrete until the slab gets its full strength. Don't wedge reshores so close to the bottom of a slab that they lift the slab and make it bow or bend upward. This can make the top of the slab crack.

Slab thickness (inches)	Design load (psf)
4	100
5	113
6	125
7	138
8	150
9	163
10	175

Figure 5-20 *Design loads for slab forms*

Metal Formwork

Builders often use all-metal or partially-metal forms on larger commercial and industrial construction projects. These include metal studs, straight and curved wall forms, wales, joists, plate girders, and trusses. You should use metal forms where you need extra strength or where you can use the same type of form again. Metal forms are more expensive than wooden forms when you only use them once, but they're more economical than wooden ones when you can use them repeatedly. You can get prefabricated metal forms for double and single wale wall forms, curved walls, columns, and battered retaining walls.

Other metal forms are made specifically for storm drainage structures such as box culverts and channels. The base of this type of form is a steel plate girder that's from 2 to 12 feet wide and from 1 to 20 feet long. These forms are for walls and roofs, and you can also use them on building construction like bank vaults, utility tunnels, and other heavy concrete structures. You can use them over again and again. They're usually called *traveling forms*, and they're mounted on wheels or moved by cranes. The basic panel is a pre-assembled unit that's bolted to a similar unit to create an entire wall or roof framework. It consists of punched steel channel outer frame, Zee ribs at 1-foot spacing, and a heavy steel faceplate. Panels are connected together with quick bolts. Economy Forms Corporation, located in Des Moines, Iowa, manufactures special forms for major structures such as concrete bridges and stadiums.

Modular Forms

Some prefabricated metal forms are made in modular form. Modular form units are prefabricated parts that include panels of various sizes, wales, studs, ties, braces, and hardware for attaching the parts together. The units are modular in the sense that they're made in multiple sizes to fit any condition. You can handset modular forms or gang them for maximum economy. The manufacturer of a modular forming system should give you installation manuals for the units.

The base of a modular form is a standard 24-inch-wide panel encased in a steel frame to protect the plywood edges from cracking or feathering or fraying. Each panel has built-in studs and walers. Side rails of two connected panels serve as a stud, and the cross-members spaced at 1-foot intervals serve as walers. Standard panels are from 3 to 8 feet long. Each length of standard 24-inch-wide panel comes with fillers 4 to 12 inches wide. There are special panels made for columns that are 26 to 36 inches wide.

You can assemble panels by inserting wedge bolts horizontally through the slots in the side rails. Then insert another wedge bolt through the slot of the first wedge bolt. To tighten the connection, tap the first wedge bolt lightly to secure the connecting panels, while pushing down on the vertical wedge bolt.

When you erect modular formwork, you can use flat wall ties or loop panel ties. Connect ties to the side rails of opposite panels and hold them in place with wedge bolts. Use braces and turnbuckles that you nail or bolt to bracing lumber connected to the forms. Other accessories include metal brackets for scaffolds, cantilevers, and brick ledges.

If you need two or more forms in a one-piece section, you can gang-form the panels. Reinforce any high sections with vertical and horizontal wood or channel walers.

Plastic Forms

A unique type of form is made of molded polystyrene and reinforced with steel mesh and rods. See Figure 5-21, which shows details from Reward Wall System, Inc. The round cavities in the form create posts and beams when filled with concrete. When completed, the form remains as part of the wall and can be finished with plaster, drywall, or siding. This type of form has several advantages:

■ There are no forms to strip, clean, transport, or store.

■ You use less labor compared to "traditional" concrete walls, wood frame, or concrete block.

■ Forms stay in place to provide a concrete wall while using less concrete than a conventional wall.

■ Forms contains built-in insulation.

■ Forms are lightweight.

■ Forms save you construction time.

Here's how to build a wall using a plastic form:

1. Pour a level foundation and install vertical rebars at 2-foot intervals.

2. Snap chalk lines on cured footing for outer edge of forms.

3. Place forms on footing along chalk lines and backfill each side a little to hold forms in place.

4. Install 4-foot long bent rebars at corners.

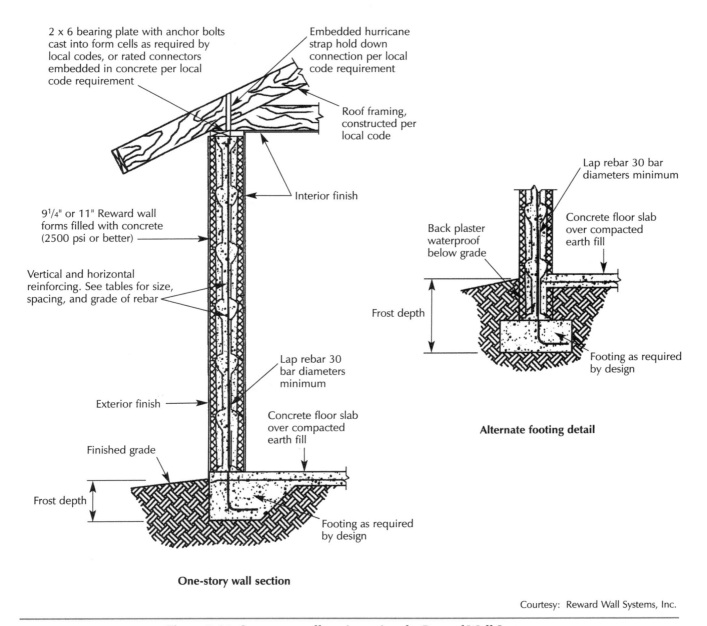

2 x 6 bearing plate with anchor bolts cast into form cells as required by local codes, or rated connectors embedded in concrete per local code requirement

Embedded hurricane strap hold down connection per local code requirement

Roof framing, constructed per local code

Interior finish

9¼" or 11" Reward wall forms filled with concrete (2500 psi or better)

Vertical and horizontal reinforcing. See tables for size, spacing, and grade of rebar

Exterior finish

Finished grade

Frost depth

Lap rebar 30 bar diameters minimum

Concrete floor slab over compacted earth fill

Footing as required by design

One-story wall section

Lap rebar 30 bar diameters minimum

Back plaster waterproof below grade

Concrete floor slab over compacted earth fill

Frost depth

Footing as required by design

Alternate footing detail

Courtesy: Reward Wall Systems, Inc.

Figure 5-21 One-story wall section using the Reward Wall System

5. Complete installation of forms, frames for window and door openings, and brace the wall.

6. Fill forms with 2,500 psi concrete with 4- to 5-inch slump.

7. Install anchor bolts at top of wall.

8. Install top plate, ceiling joists, and roof rafters.

9. Install cement plaster, drywall, or siding over wall.

Reward Wall Systems, Inc. and American Polysteel Forms manufacture this type of form.

Item	Curing time
Walls*	12 hours
Columns*	12 hours
Sides of beams and girders**	12 hours
One-way floor slabs	
Under 10-ft. span	4 days (design live load less than dead load) 3 days (design live load more than dead load)
10- to 20-ft. span	7 days (design live load less than dead load) 4 days (design live load more than dead load)
Over 20-ft. span	10 days (design live load less than dead load) 7 days (design live load more than dead load)
*Where forms also support formwork for slabs or beam soffits, use the removal time for the supported members. **Where forms may be removed without disturbing shores.	

Courtesy: American Concrete Institute

Figure 5-22 Suggested curing times for concrete

Removing Formwork

Formwork around concrete must remain in place until the concrete is self-supporting. Don't strip any forms until concrete compressive strength tests show that the concrete has reached the minimum strength required to withstand all the anticipated loads.

The minimum time to wait before you can strip forms from concrete will depend on the type of cement in the concrete. Type III high early portland cement cures much faster than Type I normal cement. If the engineer or the building code don't give you specific stripping standards, you can use the ACI (American Concrete Institute) Committee 347 suggestions in Figure 5-22 as a guide.

Renting or Buying Form Materials

You'll probably have to decide whether you should buy or rent forming equipment. In the past, formwork materials were mostly limited to boards, posts, nails, form oil, and No. 9 wire for ties. Forms were built at a job site and torn down after one use. Salvage was limited to a few pieces of timber. With rising labor costs and industrialization, mass production has become more popular. Now you can get new laborsaving and reusable materials. Distributors of mass-produced forming materials either sell or rent to builders. You can get prefabricated forms made of wood, metal, or plastic.

Here are some of the arguments for renting from a qualified, well-stocked supplier:

■ Supplier is responsible for delivery and pickup.

■ Supplier maintains backup equipment and spare parts.

■ Supplier provides design and engineering of system, such as where to place tilt-up inserts.

■ There's no leftover equipment.

■ Supplier meets short-term production peaks.

■ Supplier stocks specialty items you use sparingly or only once.

■ The material doesn't become obsolete.

■ You don't have to invest a lot to buy a forming system.

On the other hand, if you do a lot of similar formwork and have experienced crews, purchasing formwork materials may be your better choice. Figure 5-23 is a list of items you'll need to buy or rent for a poured-in-place concrete job.

Buy items	Rental items
Form coating	Curb stakes
Curing compound	Spray tank
Keyed joints	Screeding equipment
Masonite strips	Vibrators
Coil ties	Taper ties
Snap ties	Shebolts
Tie center rods	Single liner clamps
Shebolt center rods	Snap tie wedges
Fiber tube forms	Adjustable braces
Tie downs and spacers	Wale brackets
Polyethelene film	Scaffold jacks
Sisal kraft paper	Coil working parts
Cast-in place inserts	Snap tie wrenches
Panel lock bolts	Shebolt wrenches
Panel lock ties	Taper tie wrenches
Plywood hole covers	Shebolt rod puller
Expansion joint material	Coil tie wrenches
Abrasive grits	Elephant trunks
Hardener	
Curing compound	
Waterstop	
Dry-pack and grout material	
Reglets	
Anchor bolts	

Figure 5-23 *Formwork items to buy or rent*

Maintaining Forms

Inspect plywood forms for wear each time you strip them. Maintain the forms by cleaning, repairing, priming, refinishing, and treating them with a form-release agent between each use. Use a hardwood wedge and a stiff fiber brush to clean forms. You can also use acid washing for general cleaning. Wet the surface thoroughly with a 5 to 10 percent solution of muriatic acid, and scrub with a stiff bristle brush. Flush with clean water to remove the acid.

Remove all nails and fill holes with patching plaster or plastic wood filler material. You can also repair forms by applying grout and rubbing it with burlap to completely fill all pits. To remove excess grout, rub with clean burlap. Then sand the forms with No. 2 sandpaper to remove all excess mortar and make the panels smooth and uniform in color and texture. Proper maintenance will increase the life of a form and consistently produce a smooth concrete finish.

Chapter 6

Foundations

The foundation of a building is probably the most important part of the building. A building's walls, structure, and stability will depend on how accurately its foundations were built. It's probably true that a building is no better than its foundation.

Builders often use the terms *foundation* and *footings* interchangeably. When an entire foundation is in contact with soil, it's a footing. When there are two distinct parts, such as a wall, or stem wall, and a footing, you call the combination a foundation and the lower part, a footing.

Sizing a Foundation

You size a foundation to minimize subsidence and to keep different parts of the building from settling at different rates. The size is based on the bearing capacity of the soil beneath it and the amount of load on the foundation. The bearing capacity of a soil is the weight per square foot the soil can support without settling too much. Just how much is too much depends on:

- the amount of uneven or differential settlement

- how much the building can settle without permanent damage

- the amount of brittle material such as plaster and glass in the building. An all-metal industrial building with a steel frame, metal siding and roofing can tolerate more settling than a masonry or concrete building.

Most damage to a building happens when parts of the building settle at different rates from other parts of the building. To help keep a building from settling unevenly, you'd size its foundation so that each footing, whether large or small, carries the same amount of load per square foot. A unit load on a foundation is usually given in pounds per square foot, or psf. On large foundations with heavy loads, the unit load may be expressed as kips per square foot, where a kip is equal to 1,000 pounds. Loads on pilings are usually given in tons per pile.

Figure 6-1 shows various types of soil conditions and commonly-used foundations for those conditions.

Soil Testing

You can use the soil classification table published by your local building department as a preliminary guide for designing a foundation. Check Figure 6-2 for the allowable foundation and lateral bearing pressure of five types of soil, according to the 1997 *Uniform Building Code*. The 2000 *International Building Code*, in Table 1804.2, shows different allowances. Be sure you know which code you're under.

Lateral bearing pressure is the allowable soil bearing pressure, in a horizontal direction, on a vertical surface, such as the sides of a retaining wall footing. This is based on a footing that's at least 12 inches wide and 12 inches deep. To get higher allowable pressures, you make a footing deeper and wider.

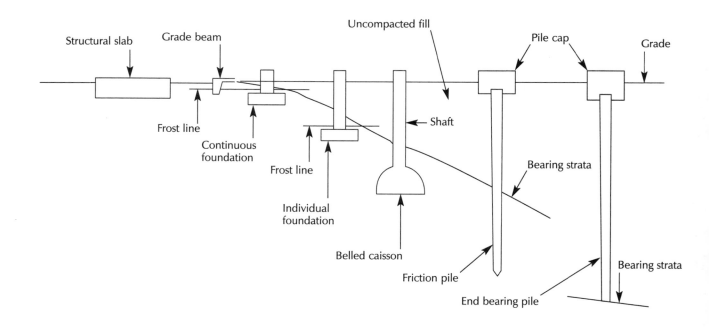

Figure 6-1 Soil conditions and foundation types

TABLE 18-1-A — ALLOWABLE FOUNDATION AND LATERAL PRESSURE

CLASS OF MATERIALS[1]	ALLOWABLE FOUNDATION PRESSURE (psf)[2] X 0.0479 for kPa	LATERAL BEARING LBS./SQ./FT./FT. OF DEPTH BELOW NATURAL GRADE[3] X 0.157 for kPa per meter	LATERAL SLIDING[4] Coefficient[5]	Resistance (psf)[6] X 0.0479 for kPa
1. Massive crystalline bedrock	4,000	1,200	0.70	
2. Sedimentary and foliated rock	2,000	400	0.35	
3. Sandy gravel and/or gravel (GW and GP)	2,000	200	0.35	
4. Sand, silty sand, clayey sand, silty gravel and clayey gravel (SW, SP, SM, SC, GM and GC)	1,500	150	0.25	
5. Clay, sandy clay, silty clay and clayey silt (CL, ML, MH and CH)	1,000[7]	100		130

[1]For soil classifications OL, OH and PT (i.e., organic clays and peat), a foundation investigation shall be required.
[2]All values of allowable foundation pressure are for footings having a minimum width of 12 inches (305 mm) and a minimum depth of 12 inches (305 mm) into natural grade. Except as in Footnote 7, an increase of 20 percent shall be allowed for each additional foot (305 mm) of width or depth to a maximum value of three times the designated value. Additionally, an increase of one third shall be permitted when considering load combinations, including wind or earthquake loads, as permitted by Section 1612.3.2.
[3]May be increased the amount of the designated value for each additional foot (305 mm) of depth to a maximum of 15 times the designated value. Isolated poles for uses such as flagpoles or signs and poles used to support buildings that are not adversely affected by a 1/2-inch (12.7 mm) motion at ground surface due to short-term lateral loads may be designed using lateral bearing values equal to two times the tabulated values.
[4]Lateral bearing and lateral sliding resistance may be combined.
[5]Coefficient to be multiplied by the dead load.
[6]Lateral sliding resistance value to be multiplied by the contact area. In no case shall the lateral sliding resistance exceed one half the dead load.
[7]No increase for width is allowed.

From the *Uniform Building Code*, ©1997, ICBO

Figure 6-2 *Allowable foundation and lateral pressure*

The best way to find out which type of foundation you need for a building is to hire a drilling rig to dig test holes at the site, and a soils engineer to make a boring log that shows the type of soil found at various depths. The soils engineer will take core samples of disturbed and undisturbed soil from the test hole, at various depths, and have them tested in a laboratory. The tests will determine the soil's bearing capacity, consolidation, moisture content, and other characteristics. The engineer will send the boring log and test results to the job architect and engineer.

You might choose to skip these tests — they're expensive and they take time. You can just take a guess about a soil type. Of course, if your guess is wrong, you'll have a financial disaster. Your building may sink or its foundation may crack. Then, you may have to underpin the foundation or remove the building and install a proper foundation.

You should follow the soil bearing values shown in your local building code for residential and commercial buildings. Industrial facilities or process plants with heavy machinery and not susceptible to plaster cracks and broken glass usually use rule-of-thumb values like those shown in Figure 6-3.

Type of soil	Bearing value (tons per SF)
Soft clay or loam	1
Ordinary clay and dry sand	2
Dry sand and dry clay	3
Hard clay and firm, coarse sand	4
Firm coarse sand and gravel	6
Shale rock	8
Hard rock	20

Figure 6-3 Allowable bearing values for common soil types

Selecting a Foundation Type

Here are the most common types of foundations:

- *Isolated foundations* are simply a footing and a pedestal. The pedestal carries the column load to the footing and the footing spreads the load to the soil.

- *Continuous foundations* are an extended concrete wall and footing. The wall carries the weight of the building to the footing and the footing transfers that load to the soil.

- *Structural mats* are thick reinforced concrete slabs underlying an entire building. The concrete mat supports the building walls and columns. The mat is very rigid, and transfers all of the loads equally to the soil. You could say that a structural mat "floats" on the soil.

- *Cantilever foundations* are special isolated foundations. They're often used at a property line where there's not enough room for a continuous foundation.

- *Caissons and piles* act as long columns carrying the weight of a building deep into the soil.

The most economical foundation you can use for a building site will depend on three things:

- loading conditions

- area available

- soil conditions

When the soil bearing value is moderate, the two most common types of foundations are the *isolated* and the *continuous* foundation. Both types are considered shallow foundations. Moderate soil is typically a sandy-loam with a carrying capacity of 1,000 to 1,500 psf. Reinforcement for a typical isolated foundation is shown in Figure 6-4. The reinforcement uses three groups of bars: a

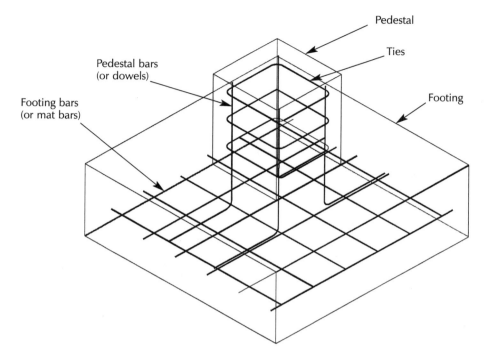

A Isometric showing reinforcement

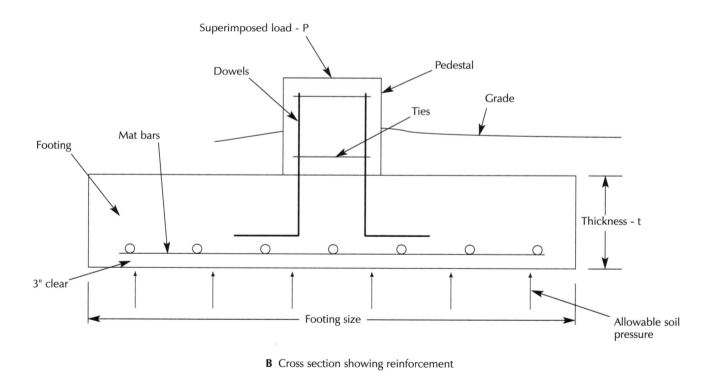

B Cross section showing reinforcement

Figure 6-4 Reinforcement in an isolated foundation

horizontal mat near the bottom of the footing, vertical bars in the pedestal, and ties around the pedestal bars.

The continuous foundation is most common type used in one- and two-story dwellings. The usual sizes and depths of these foundations are shown in Figure 6-5. If the soil is expansive, you'll need deeper footings and steel reinforcement in the foundation. For example, according to the Los Angeles City Building Code, foundations in expansive soil must be at least 24 inches below the finish grade for exterior walls, and 18 inches below for interior walls. Also, there should be at least two #4 bars, placed 4 inches from the bottom and 4 inches from the top of a foundation (§91.2905 (d) Los Angeles City Building Code).

As there are many classes of expansive soil, your building department may require soil testing and special provisions in the foundation design. Your soils engineer will recommend minimum foundation depth and maximum soil bearing value. Usually they recommend at least two continuous #4 bars in the foundation, one near the top and the other near the bottom. Table 18-1B of the 1997 *UBC* and Section 1802.3 of the 2000 *IBC* describe various types of expansive soils.

You can use the table in Figure 6-6 to figure out a preliminary sizing for a square individual column footing. Here's an example to show you how to design a square footing:

1. Find out the load. Let's say our load is 30,000 pounds.

2. Find out the allowable soil pressure. Assume the soil has an allowable soil pressure of 1,000 psf.

3. Pick a footing size from the table. Will a 6-foot × 6-foot × 10-inch footing and a 10-inch × 10-inch × 24-inch pedestal work?

4. Calculate the volume and weight of the foundation. Let's say we use concrete weighing about 150 pounds per cubic foot.

5. Calculate the total load. Converting all the dimensions of the footing and pedestal to feet, the weight of the footing and pedestal is (6' × 6' × 0.83' + 0.83' × 0.83' × 2') × 150 pcf, or 4,689 pounds.

6. Divide the total load by the area of the footing to find the actual soil pressure. The total load is 30,000 + 4,689, or 34,689 pounds. The soil pressure per square foot is 34,689 / 36, or 964 psf.

7. Now check Figure 6-6 to find out if the actual soil pressure is less than the allowable soil pressure for the footing size you picked. Since 964 psf is less than 1,000 psf and the total allowable load, 31.5 kips, is more than the 30.0 kip load, the selected footing is acceptable. As a further check, you could try calculating the soil pressure with a 7-foot square footing. The total weight of the foundation and column load is 36,945 lbs. Dividing this value by 49 square feet, the area of a 7-foot square footing, the soil pressure is 753 psf, far below an allowable pressure of 1,000 psf. The first choice is better.

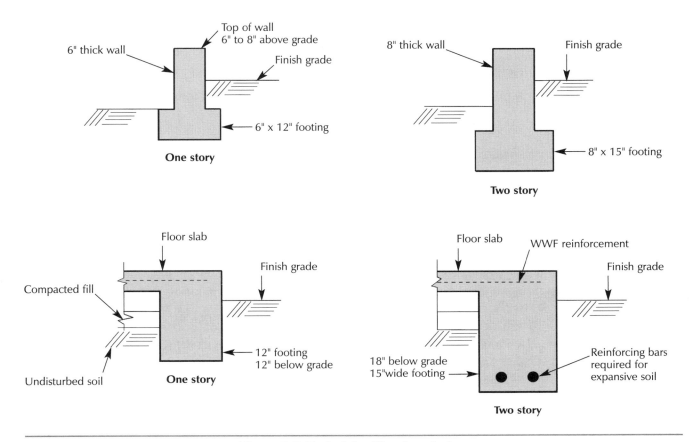

Figure 6-5 Typical dwelling foundations

Capacity (kips)	Footing size (feet)	Footing thickness (inches)	Mat bars each way
1000 psf allowable soil pressure			
21.7	5	10	#4 @ 13" o.c.
31.3	6	10	#5 @ 13" o.c.
42.3	7	11	#6 @ 13" o.c.
54.4	8	12	#6 @ 11" o.c.
67.9	9	13	#6 @ 10" o.c.
81.3	10	15	#7 @ 12" o.c.
2000 psf allowable soil pressure			
46.6	5	11	#5 @ 9" o.c.
66.2	6	13	#6 @ 11" o.c.
88.8	7	15	#6 @ 9" o.c.
115.2	8	16	#7 @ 11" o.c.
160.2	9	18	#8 @ 14" o.c.
176.3	10	19	#8 @ 11" o.c.

Figure 6-6 Safe carrying capacity of square individual column footings

If the footing wasn't adequate, you would calculate the total load again with the next size footing. Do this until you find a soil pressure that's less than the allowable pressure for the footing.

There's a foundation checklist you can use to help you build a foundation in Appendix A.

Cantilever Foundations

Cantilever foundations are special isolated foundations, often used on a property line. When a new building is close to a property line or an existing structure, a cantilever foundation may be the only type of foundation you can use. Figures 6-7 and 6-8 show two types of cantilever foundations.

The concentrated load of an exterior building column goes to the edge of the foundation, making the foundation tend to rotate. This rotation is countered by the foundation's weight, or by an interior column. When there isn't enough room for a foundation to counterbalance such a load, you need to use a cantilevered foundation. The foundation shown in Figure 6-7 uses the weight of the interior column to stabilize the load on the wall foundation and keep it from rotating. The foundation shown in Figure 6-8 uses its own dead weight to counter the load from the wall.

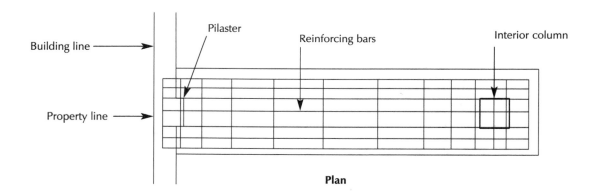

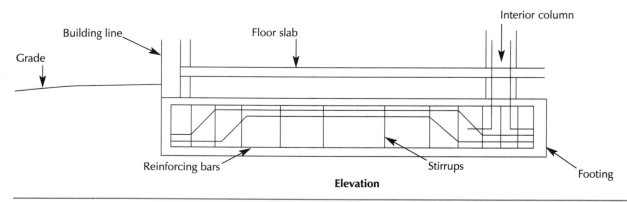

Figure 6-7 Cantilever foundation with combined footing

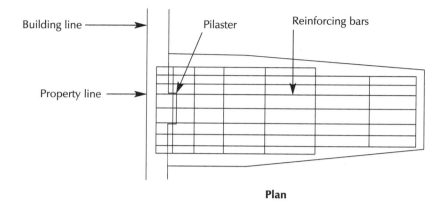

Plan

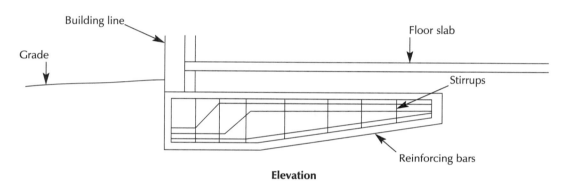

Elevation

Figure 6-8 Cantilever foundation at property line

Belled Caissons

A caisson acts as a long column that carries the weight of the building above it deep into the soil. Belled caissons are commonly used where soil conditions require deep foundations. You can also use a belled caisson foundation where there is a stiff stratum of soil under a layer of uncompacted fill.

The area of a belled caisson depends on the total of the dead and live loads on it, and the allowable soil bearing pressure at the bottom of the bell. The diameter of the shaft of a belled caisson will depend on the size of the bucket used to form the bell, the allowable soil bearing pressure of the lower stiff stratum of the earth, and the amount of load.

To make a belled caisson, drill a cylindrical hole to the depth you need. The bottom of the drilling bucket has cutting blades which rotate downward to cut a bell-shaped void in the soil. When the bucket is full, lift it up and dump the spoil material near the hole. Continue until you have expanded the hole to about twice the size of the shaft. Because of the bell shape of the hole, you rarely need to put in any reinforcement. If the top layers of soil are sandy, put a thin metal or fiber casing in the hole to stop cave-ins. After the completed hole has been inspected, use a tremie to fill the shaft and bell with concrete.

You can use caisson anchors tied to a wall with slant-drilled straps to support a retaining wall. This is effective where the property lines are so close you can't make the support excavation you need. Figure 6-9 shows a belled caisson and slant drilled strap anchoring a high concrete retaining wall. Reinforcing bars tie the strap to the wall and the belled caisson.

If you plan to excavate an area closely surrounded by property lines or adjacent buildings, you can use caissons. Figures 6-10 and 6-11 show one way to excavate, shore, and set new footings under the existing footings of a close building. Here's what you need to do at the three property lines:

1. Drill 4-foot-diameter shafts about 10 feet apart inside the property line to the depth you plan to excavate.

2. Set and brace a vertical timber, or waler, that will support the shoring.

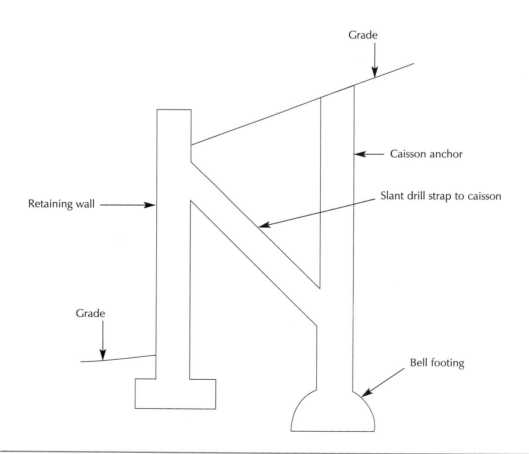

Figure 6-9 *Retaining wall anchored with strap caisson*

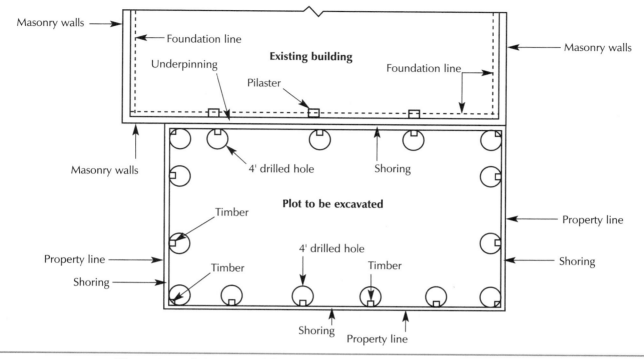

Figure 6-10 *Caissons used for excavation and underpinning foundations*

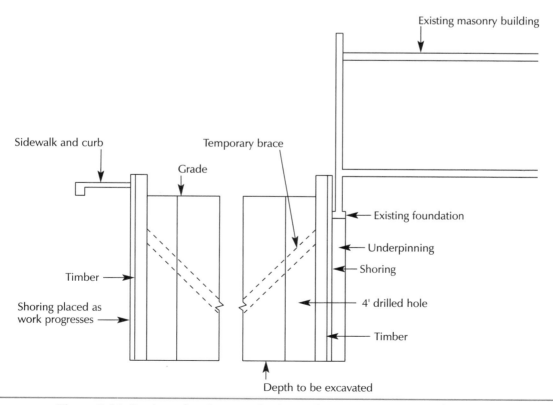

Figure 6-11 *Sections showing use of caissons for excavation and underpinning*

3. Excavate as you place individual wood shoring horizontally behind the vertical timbers to retain the earth. Shoring goes behind the vertical timbers, or wales. Shoring may also be called *planking* or *sheeting*. A bulkhead retains the fresh concrete used for underpinning, as shown on Figure 6-12.

4. Install additional temporary bracing to the vertical timbers as you excavate and place shoring.

5. When excavation is complete, build a wall form using the shoring as the outside wall form.

6. Install reinforcing bars.

7. Place concrete between the shoring and the wall form with a tremie. The back form can be left in place after the concrete sets.

Then, at the existing building on the property line:

1. Drill 4-foot-diameter shafts about 10 feet apart, or near adjacent pilaster foundations. These will serve as working chambers for the mining operation while you're undercutting 2-foot-wide slots under the adjoining building pilaster foundations.

2. Stabilize the walls of the shaft with bracing.

3. Clean all debris from the slots.

4. Bulkhead off the caisson slot and place concrete with a tremie to underpin the existing foundation.

5. Prepare to build a concrete wall adjacent to the underpinning, as on the other three property lines.

6. Set and brace vertical timbers to support the shoring.

7. Excavate as you place wood shoring horizontally behind the vertical timbers to retain the earth.

8. Install additional temporary bracing to the vertical timber as you excavate and place shoring.

9. When excavation is complete, build a wall form using the shoring as the outside wall form.

10. Install reinforcing bars between shoring and forms.

11. Place concrete between the shoring and the wall form with a tremie. The back form can be left in place after the concrete sets.

To make belled caissons for the foundation of a dock-high warehouse building, bore short belled caissons. Lay a 6-inch-thick slab to cover the compacted fill. The belled caissons support the building columns, as shown in Figure 6-13.

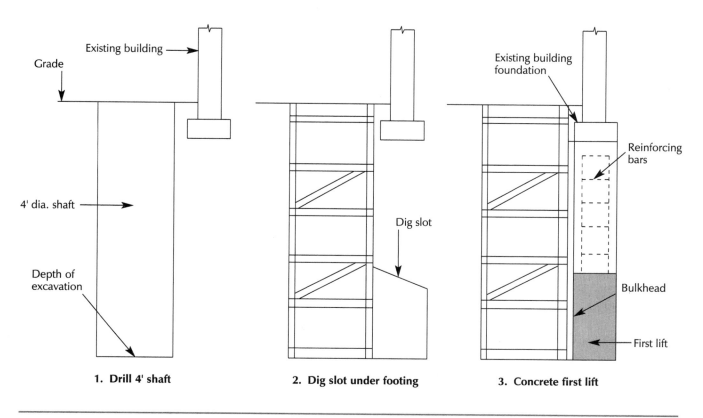

1. Drill 4' shaft　　　**2. Dig slot under footing**　　　**3. Concrete first lift**

Figure 6-12 Underpinning existing footing with drilled caissons

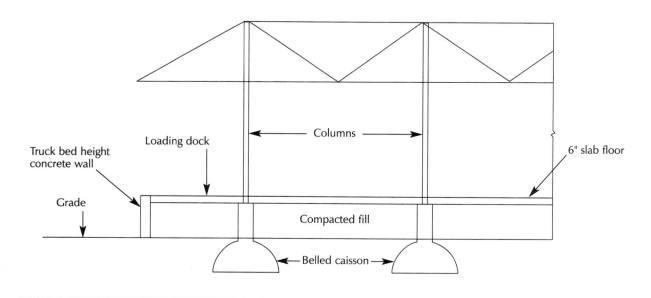

Figure 6-13 Belled caisson foundations for a dock-high warehouse

Piles

A pile is a load-carrying column that's driven into soil to support the weight of a structure. Piles can be made of treated or untreated wood, cast-in-place or precast concrete, steel, or a combination of these materials. There are two types of piles: end-bearing and friction. An end-bearing pile is a long column that extends to solid bedrock. It works like a leg on a table. A friction-type pile is more like a nail driven into wood. It uses the friction between the pile's surface and the soil to support the load on the pile. Figure 6-14 shows how loads transfer from piles to soil.

The load-carrying capacity of a pile is usually given in tons. The load-carrying capacity of a friction-type pile is generally determined in one of two ways. The drop hammer test is the most common. It counts the number of drop hammer blows it takes to drive the pile 1 foot into the ground. This number is converted into tons.

The other method is called a *static load test*. With this method, a known load is placed on an in-place test pile. Then the pile is measured to see how much it has settled after 24 hours under the load. The allowable load on a pile shouldn't be more than 50 percent of the load that will make the pile settle disproportionately. This is called the *yield point* of the pile. For example, suppose you have a pile that

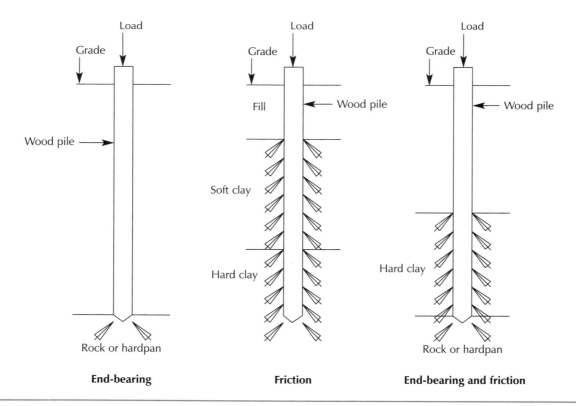

Figure 6-14 Load transfer from piles to soil

settles 0.05 inch for each additional ton of test load up to a total of 40 tons. Then, when you add more weight on the pile, the rate of settlement goes up to 0.08 inch per ton of additional load. That means that you've reached the yield point of the pile, and the allowable load on the pile is 50 percent of 40 tons, or 20 tons. The allowable load shouldn't cause a net settlement of more than 0.01 inch per ton of the test load. In addition, the amount that a pile settles should remain constant over a 24-hour period.

You can drive precast concrete, wood, or steel piles into the ground using a drop hammer or vibratory pile driver. If you use a vibratory pile driver to install a pile, you have to use the static test method to test it if your local building department requires you to test the pile.

Wood Piles

Many concrete foundations are supported by wood piles. However, wood piles can deteriorate due to decay, insect attack, and maritime-borer attack. If you bury wood piles in dry ground, only decay and insect attack can damage them. If you put the piles partially in water, all three can damage them. Decay in wood piles is caused by fungi, which need moisture and air to live. If a pile is completely in water, there's no air, so fungi can't grow. If a pile is alternately wet and dry, it may deteriorate. Treat any pile that will be alternately submerged and exposed to air to keep the wood from rotting. Usually you don't have to treat wood piles that stay submerged in water.

Metal Piles

Many concrete foundations are supported on steel piles. One popular type of pile is a metal-cased pile. Another is a combination of steel shell and concrete fill, called the step-taper pile. A closed-end corrugated steel shell is driven with a full-length steel mandrel. After reaching the proper penetration, the shell is then filled with concrete.

There are several types of steel piles that are filled with concrete. These include the *swage pile, union monotube fluted steel shell*, and the *raymond stepped steel piles*. Use the swage pile in soils where the driving is very hard and when you want to keep the shell watertight for some time before filling it with concrete. You force a light steel shell over a slightly-tapered conical precast concrete plug, insert a steel pipe, then drive the pipe with a pneumatic or steam hammer to the required depth. This causes the steel shell to swage over the plug, forming a watertight joint. Then remove the core and fill the pipe with concrete.

You can use the fluted steel shells for both end-bearing and friction load-bearing piles. To install this type of pile, drive the pile, and inspect the interior of the shell. Then fill the shell with concrete before cutting it at the proper height.

Shell sections are manufactured in basic lengths of 4, 8, 12, and 16 feet. Join shell sections to make up a required pile length. The diameter of a pile increases from the tip (bottom) to the butt (top) at the rate of 1 inch per section length. The rate of taper, or pile shape, may vary with the sections of length used. Within practical limits, you can combine many section lengths into a single pile. Tip diameters are usually 8 to 11 inches, but you can also get larger tip diameters from a pile shell fabricator.

Under normal conditions, you can go ahead and fill shells with concrete even if they're close to other piles you're driving. But watch for ground heave conditions. Heaving is an uplifting of the soil that happens when you're driving piles in incompressible clay soils, and may extend as much as several feet and several piles away. Driving piles in a group may also cause heaving. Get a soils engineer to determine which soils are susceptible to ground heave. You should take grades on each pile as it's driven.

If you drive a new pile too close to previously-driven piles, you may cause the previously-driven piles to move out of position. If the pile shows signs of rising, redrive it unless it's a friction pile in a uniform stratum. Where heave occurs, leave shells unfilled until pile driving has progressed beyond the heave range and you've redone any piles that need to be driven again. Take care not to lose the grade markings if you cut off the pile before completing driving the group.

You can use step-taper piles for most types of soils, and as both friction and end-bearing piles. Usually, these piles are supported through a combination of friction and direct bearing. The tapered shape of this type of pile gives it higher load-carrying capacities than similar piles that don't have a taper. Internal steel reinforcement is normally not required in step-taper piles or caissons. However if the piles are subject to uplift, high loads, or they extend through air, water, or very fluid soils, you should consider reinforcing them.

Cast-in-Place Piles

You make cast-in-place piles by placing concrete in a drilled hole without using any forms. There are special techniques and design mixes for placing concrete into piles or caissons to make sure the pile shell is filled completely and uniformly. The most common way is to use a tremie. Lift the tremie slowly to let the concrete flow out. You should use a hose to carry the concrete to the bottom of the casing. The pipe must be long enough to reach from the working platform or ground to the lowest point you're pouring concrete to. You should keep the bottom of the discharge end buried in the newly-placed concrete the entire time you pour the concrete. Pumping from a mixer is the best way of supplying concrete. You can suspend a tremie by a crane boom and use the boom to raise and lower the tremie.

The concrete you use should have a slump of 6 inches. About 60 percent of the total aggregate should be sand. The maximum size of the coarse aggregate should be $1^1/_2$ to 2 inches. Use a plasticizing agent to give the concrete a uniform consistency.

Soil and Design Considerations for Piles

Before you select any type of pile foundation, carefully check the properties of the soil you'll be working in. Here are some of the questions you should consider:

1. What is the depth of the bearing stratum? The bearing stratum is the top of the natural soil that is capable of supporting the weight of a building, such as the bottom of an uncompacted fill.

2. What is the friction factor of the soil? The friction factor of a soil is the amount of resistance the soil has to the surface of a pile. The total load-carrying capacity of a friction pile is equal to the friction factor of the soil times the total surface area of the pile.

3. Is there a fill that may make the piles sink? If you drive a pile through a sinking or subsiding fill, the fill can drag the pile down.

4. Do you need batter piles for wind or earthquake loads? Batter piles are piles you drive into the ground at an angle rather than vertically.

5. What are the expected loads on the piles, including dead load, live load, and load from wind or earthquake?

6. How many piles are required in each group? When the total column load is greater than the carrying capacity of an individual pile, you should add additional piles in a cluster. For example, if the total building column load is 60 tons and the carrying capacity of a pile is 20 tons, use a three-pile cluster.

7. What is the minimum spacing between piles? Normally you don't place piles less than 3 feet, or 3 pile diameters, apart, whichever is greater. Friction piles are usually tapered outward from the tip to the butt. Using the diameter of the tip and the amount of taper for a specific type of pile (wood, precast concrete, or metal casing), you can use the pile's length to find the butt diameter. For example, if a pile's diameter increases by 1 inch in 10 feet from the tip to the butt end and the tip is 8 inches in diameter, a 30-foot pile would have an 11-inch diameter butt [8 + (30/10) = 11].

    The number of piles and their diameters govern the shape and size of the pile cap. A pile cap is similar to a footing, except that it's supported by the top of a pile cluster, not soil. Generally, set the piles 24 to 30 inches on center, or leave 12 inches clear between piles. Place the piles symmetrically around the column or pilaster. Check with your local building department for other requirements.

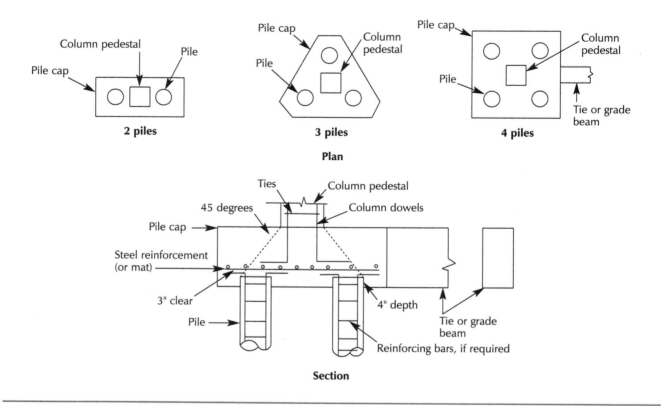

Figure 6-15 *Typical pile caps*

8. What is the minimum depth of the pile cap above the top of a pile? Usually you set the top of a pile at least 4 inches into the bottom of a pile cap. The thickness of the pile cap above the top of a pile is determined by a 45-degree plane between the pile and the pedestal on the pile cap. The steel reinforcing mat at the bottom of a pile cap is at least 3 inches above the top of the pile. When you have all this information, you'll know the number, length, and diameter of piles and the size of the pile cap. Figure 6-15 shows three kinds of pile caps: clusters of two, three, and four piles. Tie the pile cap together with the grade beams or tie beams. Design these ties to withstand a tensile or compressive load equal to 10 percent of the maximum column load on the pile cap or caisson. A typical concrete tie is 12 inches square, with four #6 bars and #3 ties at 12 inches on center.

Restrictions on Adjoining Excavations

A common construction situation occurs when you build a large multistory building with underground parking adjacent to an existing one-story building. The new construction can cause undermining or loss of support to the existing shallow foundation adjacent to the property line.

Figure 6-16 shows a cross-section of an existing building near the property line and a proposed excavation. If you're going to excavate 12 feet or less, you should protect the excavation so the soil of the adjoining property doesn't cave in or settle. But according to the *UBC*, you're not liable for the expense of underpinning or extending the foundation of the adjoining building. Some codes use a 10-foot depth instead of 12 feet. Check with your local building department.

If you're going to excavate more than 12 feet, you should protect the excavation as described above, and extend the foundation below 12 feet of depth. The owner of adjoining property is liable for the cost of underpinning when the excavation is less than 12 feet below grade, and you, the builder, are responsible for the rest of the extension of the foundation.

Figure 6-17 shows one way of protecting the excavation. You drive sheet piling adjacent to the property line before excavating. You can underpin an existing foundation by excavating and casting concrete under a 3-foot length and full width of the footing, then skipping 6 feet. After the concrete has set up, excavate and place another 3-foot length of foundation and repeat again after the concrete has set. You will now have a continuous underpinning that carries the building load to the soil.

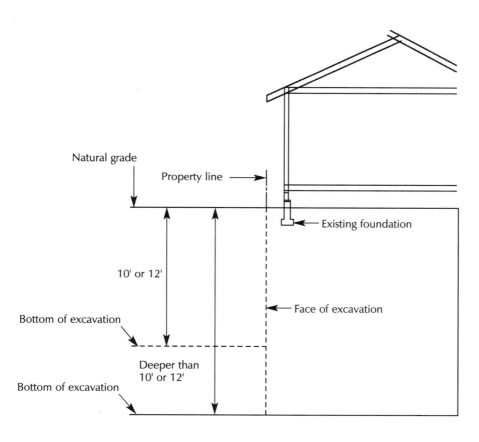

Figure 6-16 Excavation adjacent to property line

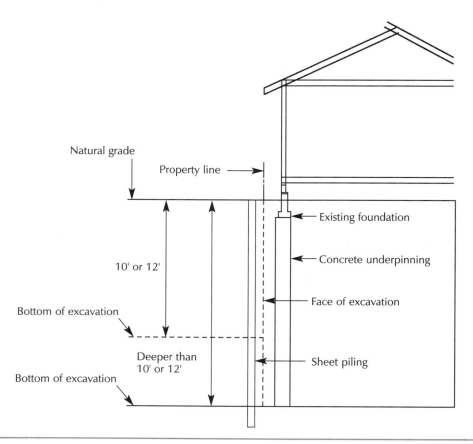

Figure 6-17 Underpinning and sheet piling

Rules for excavation adjacent to a property line are included in the *Standard Building Code*, the BOCA *National Building Code*, the *Uniform Building Code* and the new 2000 *International Building Code*. These rules provide information about how close and how deep you can excavate next to another building, and when and how to notify the owner of an adjacent building. In the 1997 *Uniform Building Code*, Section 3301.2 *Protection of Adjoining Property* says you must notify the owners of adjoining buildings in writing at least ten days beforehand that the excavation is to be made and that their buildings should be protected. You also have to give the owners of the adjoining properties access to the excavation for the purpose of protecting their buildings. The 2000 *International Building Code*, in Section 3307, says much the same thing, though it omits the owner-access requirement.

Grade Beams

Grade beams are the rectangular concrete members placed around the outside of a building. They support the exterior walls. The two lower footings shown back in Figure 6-5 show a grade beam and floor slab cast together. This is called *monolithic* construction. The upper two footing details show a continuous foundation. A continuous foundation for a building with a slab floor is sometimes called a grade beam because the foundation acts like a beam at grade level.

Another type of grade beam is a reinforced concrete beam cast at grade level that ties pile caps or caissons in position. These beams distribute wind and earthquake forces over the entire foundation system.

Foundations for Non-Residential Buildings

You have to be more precise building a non-residential foundation than building a residential one. If you make an error placing anchor bolts in a dwelling, you probably won't have any dire consequences. But if you go wrong placing anchor bolts in a commercial high-rise building or a process plant, you can run up some large costs fixing the problem.

High-rise commercial buildings are usually framed in steel. Column base plates are shop-welded and drilled. To assure that the anchor bolts are set correctly, the steel fabricator will supply loose leveling plates with drilled holes. Use these as a template so you can accurately place all the anchor bolts. You set the height of the leveling plate with a surveying instrument and hold it in place with nuts attached to the anchor bolts. Then pack the gap under the leveling plate with non-shrinking grout.

You have to be especially careful with any industrial construction work you do. Industrial equipment is usually large and expensive. And equipment such as rock crushers, rolling mills and compressors are always vibrating. You must anchor these types of equipment solidly to an adequate foundation. A very large machine can move if its anchor bolts are too small or its foundation is too light. You don't want a machine going out of balance and walking away from its foundation. The manufacturer of a piece of equipment will specify the required size and location of its anchor bolts. Some machines rest on cast iron bases with machined holes that are only $1/16$ inch oversize. That means that you can't be off more than $1/16$ inch off placing the anchor bolts for those machines.

Although industrial work includes buildings, the most complex foundations are for vessels and machinery. You won't find guidelines in the building codes for designing or constructing this type of foundation. As a rule of thumb, the weight of a block foundation for a rotary machine should be at least three times the weight of the machine. A reciprocating machine has a greater dynamic motion than a rotary machine so it requires a foundation that weighs at least five times its weight. Typical examples of reciprocating machines are piston-driven compressors and gasoline engines. Rotary machines include centrifugal pumps and electric motors. It's also a good idea to install vibration isolators between a foundation and a dynamic machine. You can use neoprene pads, ductile iron pads and air cushions, or spring-type isolators.

Equipment Plot Plan

A foundation location plan is a plot plan that shows both above-ground and underground foundations and footings of process equipment, pumps, buildings, pipe supports, etc. You should draw this plan in as large a scale as possible, usually 1 inch equals 10 feet. Show the item numbers and location of all equipment. Locate a vertical vessel by the intersection of its centerlines. Show horizontal vessels, pumps, and compressors by their longitudinal centerlines and end anchor bolts. Show concrete notes on a plot plan. A foundation location plan serves many useful purposes. If there are overlapping foundations or interference between pipelines and underground foundations and footings, they should show up on the foundation location plan. The plan should show the extremities of the underground concrete. To resolve any problems, you can combine the footings. Or you can spread equipment further apart so the footings don't touch.

Figure 6-18 shows a typical foundation location plan for a particular part of a process plant. This one includes horizontal and vertical vessels, pumps, and pipe supports. The various types of equipment are indicated by letters, like V-1, V-2, and P-1. This plan locates each item by the coordinates of the equipment's centerlines. Other foundation location plans may include a cooling tower area,

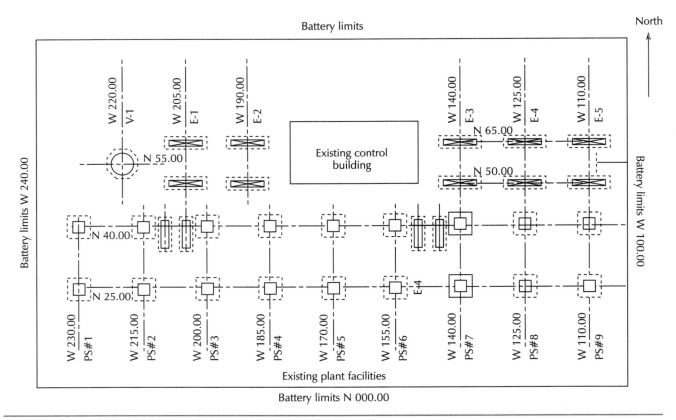

Figure 6-18 *Foundation location plan for a process plant*

furnace areas, and so on. These plans differ from a plot plan in that they show the footings below grade and foundations above grade. Use separate drawings that describe how to build each type of foundation.

Usually, foundation location plans and plot plans include General Notes specific for those plans. If the information isn't shown on these plans or in the General Notes, use the job specifications. Information not shown may be noted elsewhere on the plans where similar conditions exist. For example, if there are three identical pumps, but only one pump foundation has dimensions noted, use these dimensions for all three pump foundations.

Here are some items you should include in the General Notes section of a plot plan:

- List references to other drawings.
- Dimensions, unless otherwise noted, are to rough concrete finish, centerlines of column footings, walls, and building lines.
- All reinforcement must be deformed bars conforming to ASTM A-305.
- All wire mesh must conform to ASTM A-185.

Foundations for Process Units

There are generally six major types of foundations for process units. They are:

- Square footing with a pedestal, usually used for a vertical vessel under 3 feet in diameter.
- Octagonal footing with an octagonal pedestal, usually used for vertical vessels over 3 feet in diameter.
- Octagonal pier, usually used for a small vertical vessel.
- Rectangular block, used for pumps, compressors, and similar equipment.
- Rectangular mat, used under a group of pumps, compressors, and similar equipment.
- Wall and footing foundation, used to support horizontal vessels, heat exchangers, and reboilers. Vessel diameters can vary from 2 to 12 feet.

Horizontal Vessel Foundations

Horizontal vessels include heat exchangers, reboilers, accumulators, propane storage tanks, and similar equipment. You can place exchangers side-by-side or stack one on top of the other. It's important to know which end of an exchanger is the channel end and which is the shell end. The internal tubes are pulled from the channel end so you have to anchor the shell end securely to a foundation pier. Because a vessel will expand and contract, you should mount the saddle (the curved part of the two walls that support the horizontal vessel) on a steel plate that lets the vessel slide. Figure 6-19 shows a typical horizontal vessel and its foundation.

Here are some recommendations for designing a horizontal vessel foundation:

■ Make both piers the same size.

■ Minimum pier size is the saddle dimension plus 4 inches (both ways). This makes a 1-inch clearance all around the concrete.

■ Make each footing 6 inches wider than the width of its pier. Make the footings at least 24 inches thick.

■ For piers 10 or more inches thick, use a double layer of reinforcement on each side.

Figure 6-20 shows the thickness of the supporting walls of a horizontal vessel foundation and the size and spacing of the horizontal and vertical reinforcement recommended for the walls.

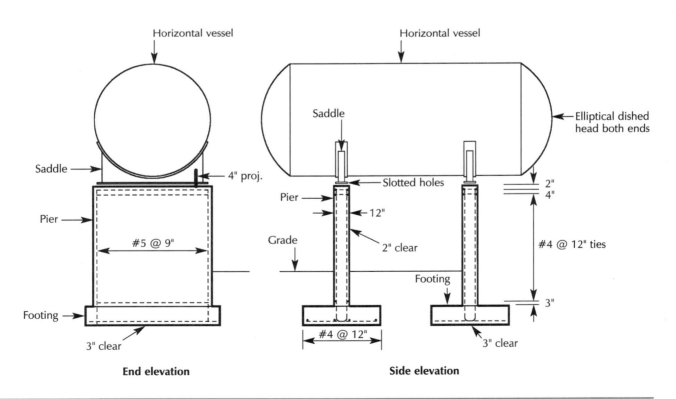

Figure 6-19 *Typical horizontal vessel and foundation*

Supporting wall thickness	Vertical bar reinforcement size and spacing	Horizontal bar reinforcement size and spacing
8"	#4 @ 9"	#3 @ 12"
10"	#4 @ 9"	#3 @ 12"
12"	#5 @ 9"	#4 @ 12"
14"	#5 @ 9"	#4 @ 12"

Figure 6-20 *Supporting wall thickness and recommended reinforcement for a horizontal vessel foundation*

Vertical Vessel Foundations

Some common vertical vessels are bubble towers, fractionating towers, and reactors. Typically, they are 3 to 12 feet in diameter and up to 100 feet in height. You should support these vertical vessels, at grade, on concrete pedestals built over rectangular or octagonal footings. An octagonal foundation is an eight-sided foundation (shown in Figure 6-21) that distributes its load over a large area. You can use rectangular footings for vessels under 4 feet in diameter. Use a rectangular footing for vessels supported on angle legs.

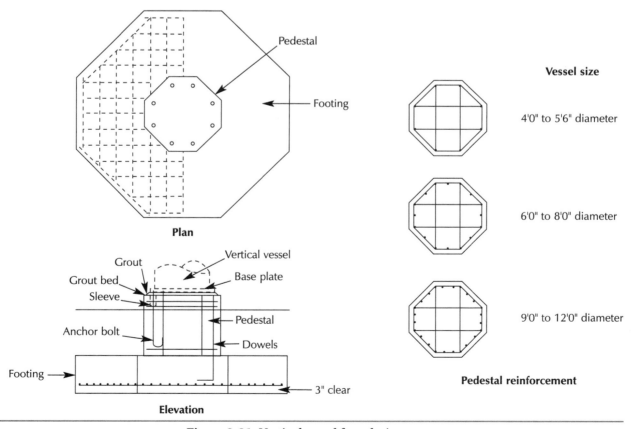

Figure 6-21 *Vertical vessel foundation*

The pedestal (concrete base over a footing) rises above grade or pavement to form the exposed direct contact surface for the vessel it supports. The base plate, which is the lowest part of a vessel, is a flat plate welded to the bottom of the vessel support or skirt. It rests on the concrete pedestal. The pedestal should be in 6-inch increments, and wide enough so there's a 4-inch clearance between the steel anchor bolts and the face of the concrete. Or it should be wide enough so there's a $2^1/_2$-inch clearance between the steel anchor bolts and the edge of the concrete.

Make footings at least 12 inches thick. Here are some other reinforcement requirements for foundations of vertical vessels:

- For 4'0"- to 5'6"-diameter vessels, use #4 ties at 15 inches on center, maximum, and four #4 U-bars.

- For 6'0"- to 8'6"-diameter vessels, use #4 ties at 15 inches on center, maximum, four #5 U-bars, and eight #5 L-bars between U-bars.

- For 9'0"- to 12'0"-diameter vessels, use #5 ties at 15 inches on center, maximum, four #5 U-bars, and sixteen #5 L-bars between U-bars, four #4 U-bars.

You should set anchor bolts no less than 3 inches from the bottom of a footing.

Flat Bottom Tank Foundations

If you want to set a flat bottom tank on poor soil, it's recommended that you lay an 8-inch-thick reinforced concrete slab for the tank first. Extend the slab about 3 inches beyond the outer edge of the tank bottom plate. Surround the tank with a concrete wall that's at least 30 inches deep. Project the wall about 4 inches above ground level. Reinforce the wall with steel bars that are about 0.25 percent of the cross sectional area of the wall. Place asphalt flashing between the tank and the wall at ground level. If the soil is compacted, use an 8-inch-thick sand or gravel bed in place of the concrete slab. Compact the sand or gravel and cover it with asphalt saturated felt (roofing paper). Build a concrete ring wall as described above.

If the soil is very good, you can omit the concrete slab. Place a sand cushion of at least 6 inches of level sand instead of the slab. It's good practice to saturate the sand with discarded oil and add moistened and compacted gravel after removing any uncompacted surface. Install a concrete wall as described above.

Pump Foundations

There are generally two types of pump foundations: rotary (or centrifugal) and reciprocating. A reciprocating pump foundation must have a depression on its top with a pipe drain. Both types need a grout bed 1 to $1^1/_2$ inches thick. Most pumps have integral motors, but large pumps may be driven through a gearbox. In this case, make the foundation rigid enough to maintain correct alignment with connected machinery. The foundation should have a flat mounting surface to support the unit uniformly. Allow the concrete to set up firm before bolting a unit to a foundation.

You should set all tanks, vessels, heat exchangers, heaters, reboilers and other such equipment on a nonshrinking grout bed. The bed may be 1 to 4 inches thick, depending on the size and weight of the equipment.

Figure 6-22 shows typical foundations for a rotary and reciprocating pump. Use similar foundations for other dynamic machinery. The weight of the foundation for rotary machinery should be at least three times the weight of the equipment, and five times for reciprocating machinery. Remember to provide anchor bolts, sleeves, electrical grounds for the motors, and drains for steam-driven pumps.

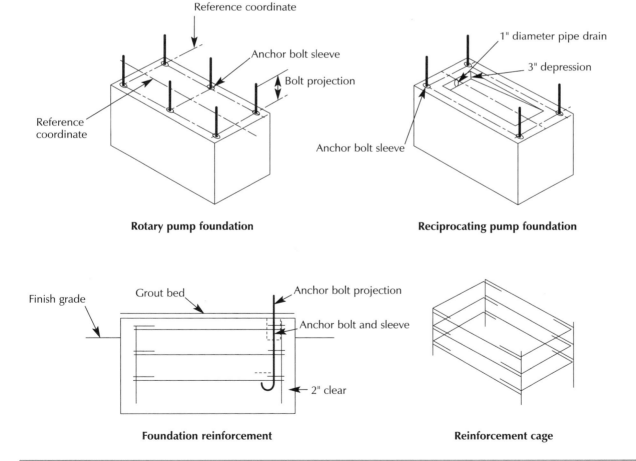

Rotary pump foundation

Reciprocating pump foundation

Foundation reinforcement

Reinforcement cage

Figure 6-22 Typical pump foundations

Anchor Bolts

Anchor bolts are made from round steel rods with a J or L shape at one end, which you embed in the concrete. The upper end is threaded. Some anchor bolts have sleeve nuts. Figure 6-23 shows three types of anchor bolts. Foundation detail plans usually include anchor bolt schedules or tables that describe the type, dimensions, and quality of material of all the bolts and sleeves. The letters are shown on a table for each diameter anchor bolt, and indicate the following:

A = Distance from the top of the threaded end of the anchor bolt to the bottom of the nuts

B = Diameter of the sleeve

C = Length of the sleeve

D = Length of the lower section of the anchor bolt

E = Length of the straight portion of the anchor bolt

F = Total depth of a J-type anchor bolt

G = Distance to the beginning of the hook

H = Total depth of embedment of a J-type anchor bolt

d = Diameter of the anchor bolt

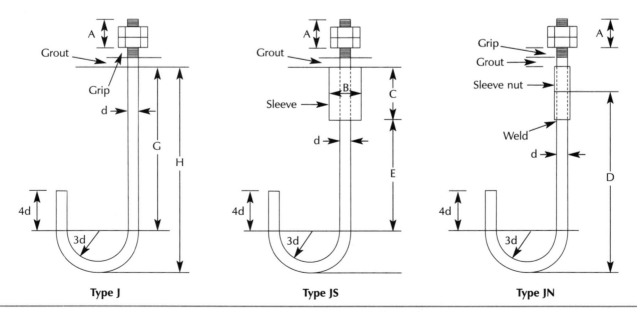

Figure 6-23 *Foundation anchor bolts*

Anchor bolts are very important in industrial facilities. They must be the right size and in the correct location for proper installation of very expensive equipment. The manufacturer of a piece of equipment should give you a certified drawing that shows the required size, location, and projections of the anchor bolts required for the equipment. This drawing is more important than the engineer's drawings or catalog illustrations because the bolt size and location on it are guaranteed (or certified) by the manufacturer to be correct. The engineer's plans may have been based on a sales brochure. If the certified plan doesn't match the engineer's plan, call the engineer immediately and get the mismatch resolved.

Even if the certified plan does match the engineer's drawings, it's still a good idea to use anchor bolt sleeves. Use a pipe sleeve 3 or 4 inches larger than an anchor bolt so the sleeve surrounds the bolt. Make sure the pipe sleeve is held solidly in place by concrete. You can move a bolt slightly however, to the corresponding hole in a pump's base plate. Use a grout bed made of a rich mixture of cement, sand, and water to level and form the elevation of equipment. Anchor bolt sleeves give you more flexibility in setting the bolt location.

The sleeve nut extends the length of a bolt. Sometimes, you can use machine bolts instead of anchor bolts for light machinery. Anchor bolts have a rougher finish than machine bolts and conform to ASTM A-307. They usually come with square nuts, while machine bolts have hex nuts. Anchor bolts are usually embedded in concrete. Machine bolts are made with higher precision and are used mainly to fasten machine parts. Common machine bolts conform to ASA B18.2 and High Strength machine bolts conform to ASTM A325.

The structural steel fabricator should furnish you with anchor bolts and leveling plates. If they don't furnish a leveling plate, they should furnish a template for setting the anchor bolts accurately.

Smaller pumps may be anchored with cinch anchors, bolts that self-lock in drilled holes by expanding in concrete after the concrete is hard. Use the L-shaped bolt to anchor moderate loads. It's similar to the type you use on mudsills in dwelling foundations. Use the J-shaped bolt to anchor heavy loads. (Moderate loads include static building parts, like mudsills. Heavy loads include dynamic machinery like pumps, or loads with an uplifting force.)

Here are some more recommendations for sizing machinery anchor bolts:

- Use "J" or "JS" anchor bolts because they have greater holding capacity than the L-type anchor bolt.
- Maximum unit tensile strength of a bolt is 15,000 psi without allowance for corrosion.
- The minimum clear distance from a sleeve to the edge of the concrete is 4 inches. A sleeve should never be less than $2^1/2$ inches from the face of concrete.
- Use sleeves for vertical vessels, pumps, compressors, and equipment with cast iron bases.

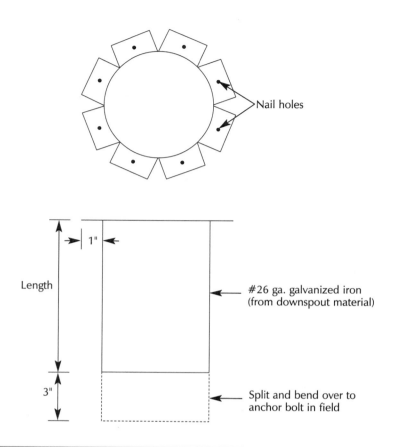

Figure 6-24 Anchor bolt sleeve

- Use two nuts per bolt for vertical and horizontal vessels, shell and tube-type exchangers, and reciprocating machinery.

- Find the projection from the threaded end of the anchor bolt to the top of the concrete. It should be long enough to accommodate one or two nuts, a base plate, a leveling plate (if any), the grout bed and at least $1/4$ inch extra.

- Where you use slide plates for thermal expansion, place jam nuts on anchor bolts and slide plates to provide at least a $1/16$-inch clearance between the lower nut and the base.

You should place anchor bolts for process equipment within $1/16$ inch of the location specified on the plans if possible. As a safety measure, set each anchor bolt in a sleeve so you can bend the bolt to fit it in the holes in the base plate or equipment. Make sleeves of #26 gauge galvanized steel (same as roof downspout material). The diameter of a sleeve varies from 3 to 6 inches, and the length varies from 6 to 12 inches. See Figure 6-24. Cut 1-inch-long slits at the top end of the sleeve and bend the tabs horizontally to form a flange. Split 3 inches from the bottom end. Then bend the sleeve's tabs inward toward the anchor bolt. Be careful while you pour concrete. Don't fill the sleeves with concrete. Pack them with plastic, rags, or paper before you pour.

Chapter 7

Floor and Roof Slabs

Concrete floor and roof slabs are important parts of a building. In fact, you design an entire building based on the purpose of its floors. You can design elevated slabs (slabs not cast on grade) as flat plates, flat slabs with dropped panels, one-way or two-way solid slabs, pan joists, and several other variations.

Reinforce slabs with steel bars or wire fabric. Reinforcing bars serve two purposes. *Load bars* resist tensile stress in the slab from loads that could bend it. Since concrete doesn't resist tensile stress well, the steel bars do that work. *Temperature bars* are used to control shrinkage due to temperature change. When a freshly-cast concrete slab cures, it shrinks. Concrete also shrinks when air temperature changes from hot to cold. This can make concrete crack unless there's steel reinforcement holding the concrete together.

The minimum thickness of a slab is $3^1/_2$ inches, although it's usually specified as a 4-inch nominal thickness. That's because you use a 2 × 4 as a screed, and a 2 × 4 is only $3^1/_2$ inches wide.

Floor Slab Defects

Some cracks in a slab are inevitable. These are some of the more common types of cracks:

■ *Settlement cracks*. In slabs on grade, these may be due to a weak subgrade.

- *Drying shrinkage cracks.* These appear when water sheen disappears from the surface. They're usually random, straight, hairline cracks that extend to the perimeter of a slab.

- *Plastic shrinkage cracks.* These are wider than drying shrinkage cracks and they extend through the depth of a slab. They're parallel to each other, and perpendicular to the wind. They occur mostly on dry, windy days. Apply a fine spray of water to help close them or reduce their penetration.

- *Shifting form cracks.* These are caused by wood expansion, or nails or clamps loosening. If you place reinforcing steel too close to the surface of a slab, you may get cracks over the bars as the concrete settles. Also, rusting reinforcing bars will expand and cause cracking if there's not enough concrete cover over them. An open crack lets more water reach the bar and increases the rate of rusting. To prevent this, allow $1^{1}/_{2}$-inch cover over the bars and use a fairly low-slump concrete. Low-slump concrete is stiffer than moderate-slump concrete, as there's less water in the mixture so it shrinks less. High-slump concrete shrinks more as the excess water dries out, causing contraction cracks in the slab.

Here are some examples of things that'll make a slab discolor, craze, or scale:

- Dusting a different brand or type of dry cement over a wet cement surface may cause differences in cement color.

- Over-troweling may cause scaling and burned areas. If you trowel concrete too much, you'll get a burned, or burnished, surface that's very smooth and has a high gloss finish. This is caused by the fine aggregate and cement rising to the surface while the larger aggregate is forced down. This type of surface is not recommended for slabs subject to foot traffic as the slab becomes slippery when wet.

- Using different slump concrete in different areas will change colored concrete.

- Changing the brand of cement you use in a concrete area may cause differences in cement color.

- Adding calcium chloride to a mix affects concrete color.

- Placing concrete on a wet subgrade that has water puddles lets water leach out through the concrete to the surface. That can leave that area lighter in color.

- Letting concrete dry too quickly on low humidity days and/or wind blowing across the surface can cause crazing. You can correct this by grinding.

- Letting concrete get too cold so it doesn't cure properly. This can result in crazing and scaling.

- Low air content can cause scaling.

- Using too many salts or deicing chemicals can cause scaling.

- Using high slump concrete can cause scaling.

- Using an improper curing method can cause scaling.

The following are remedies for some of these defects:

- Don't use a concrete slump over 4 inches.

- Use 4 to 6 percent entrained air in concrete.

- Don't trowel while bleed water is present.

- Don't over-trowel.

- Slope concrete for drainage.

- Allow adequate curing time.

Slabs on Grade

Slabs you cast on grade need special attention to prevent cracking or moisture getting into the slab. Embed welded wire fabric in concrete to help keep it from cracking during its curing phase. To be effective, the area of steel in a concrete slab should be at least 0.2 percent of the area of the slab. For example, the cross-sectional area of a 6-inch-thick slab 12 inches wide is 72 square inches. The minimum reinforcement is 0.002×72, or 0.144 square inches. You can also use a mat of welded reinforcing bars instead of wire mesh.

If there's moisture in the subgrade, an interior slab may act like a blotter, drawing water to the surface. So, you should cast interior slabs over a bed of coarse aggregate to keep water from penetrating into a slab. It's also a good idea to install a waterproof membrane under the slab.

Joints in Slabs

To reduce or control cracking due to temperature changes, install special types of joints such as expansion joints, weakened-plane joints, control joints, or construction joints. You use these joints on both slabs on grade and elevated slabs.

You should design *expansion joints* as a tongue-and-groove (T&G) connection. Adjacent slabs should be able to expand and contract freely, but should lock into each other to keep them from moving vertically. Provide a gap of about $1/2$ inch between the tongue and the groove and fill the gap with a waterproof caulking. To form the T&G bulkhead, you can use wooden forms with a beveled strip, or use preformed metal forms. T&G joints are also called key joints.

Weakened-plane joints, or scoring joints, are commonly used in public sidewalks. They should be transverse to the line of work and at regular intervals of about 5 feet, but always less than 10 feet. After preliminary troweling, you part the concrete to a 2-inch depth with a straightedge to create a division in the coarse aggregate. Use a jointer tool with a depth of $1/2$ inch and radius of $1/8$ inch. Refloat the concrete to fill the parted joint with mortar. Mark the edge forms to locate the 2-inch-deep groove in the slab. Don't make the finish joint opening wider than $1/8$ inch.

Place *control joints* in floor slabs to reduce random cracking and to predetermine where cracks will occur. The depth of a control joint depends on the thickness of the slab. Make the control joint while you're finishing the concrete. Or, you can cut it after the concrete has set at least 12 hours. To install a plastic control joint, use a T-shaped strip at least 1 inch deep with a suitable anchorage and stiffener. This is to prevent vertical movement. After preliminary troweling, part the concrete 2 inches deep with a straightedge. Insert the plastic strip in the impression so that the upper surface of the removable stiffener is flush with the concrete.

Construction joints are joints between two separate pours. This type of joint is similar to an expansion joint except that there's no gap between the pours. You can also use $1/2$-inch round steel dowels to hold the two slabs together.

Here are some other things you can do to control cracking:

■ Reduce the amount of water in a concrete mix.

■ Distribute the stresses so the cracks won't be noticeable or damaging. You can distribute the stresses due to shrinkage by dividing the slab into smaller sections with construction joints. For example, divide an L-shaped or T-shaped pavement with construction joints so that you'll only have rectangular shapes.

■ Confine cracks to predetermined, unobjectionable locations, as shown in Figure 7-1.

Sizing an Elevated Slab

To make a preliminary sizing of an elevated slab in a residential building, you can figure that the live load is 40 pounds per square foot. A live load is any temporary imposed load such as occupants, appliances, or furniture. In most areas, the minimum live load for roof slabs is 20 pounds per square foot. In localities where snow can build up, the live load may be increased to 30 pounds per square foot. In any case, check with your local building department. The live load for a residential garage is usually 50 pounds per square foot.

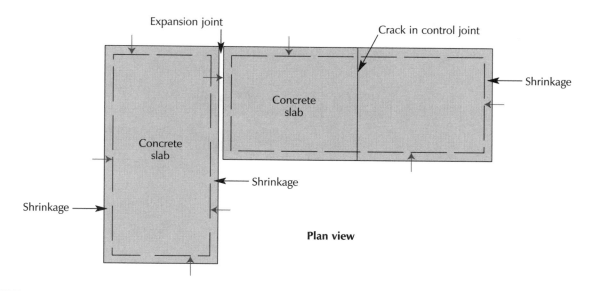

Plan view

Figure 7-1 Concrete slab with expansion and control joints

Every slab must also carry a dead load. The dead load is the weight of all permanent structural and nonstructural components. This may include ceilings, partitions, piping, and flooring. When an elevated slab is part of an office building, the slab must carry all other dead loads in addition to its own weight. This may include the suspended ceiling, ductwork, sprinkler piping, plumbing, permanent partitions and walls, and flooring. Movable partitions are considered to be part of the live load. On warehouse or storage building floor slabs, the dead load would only be the slab. Stored merchandise and lift trucks are considered to be live load.

Here's a checklist you need to consider when you select any slab that you don't cast on grade:

1. How will the slab be used?

2. What are the loads on the slab — light or heavy, uniform or concentrated?

3. Is the span of the slab long or short?

4. How will the slab be supported?

5. Are there any obstructions to the columns, beams, and girders?

One-Way Slabs

The simplest and most common type of concrete slab is the one-way slab. A one-way slab spans from beam to beam, or from wall to wall. The beams may be steel, masonry, or concrete. The load-carrying reinforcement is placed near the bottom of the slab, perpendicular to the direction of the beams. Temperature reinforcement, which doesn't carry any load, is positioned near the top of the slab parallel to the beams.

There are two types of temperature reinforcement you can use: deformed steel bars and welded wire fabric. The minimum amount of temperature steel is determined by the thickness of the slab. Figure 7-2 lists the typical reinforcement for slabs that are 2 to 8 inches thick. Figure 7-3 shows typical reinforcement for one-way floor slabs in residential buildings.

A cantilevered concrete balcony is an example of a one-way slab. The load bars project out from the floor slab and are placed near the top of the slab. You place temperature bars parallel to the building line and under the load bars. Although cantilevered concrete balconies are architecturally attractive, they can spall and crack if not constructed properly.

Moisture can penetrate inadequate concrete cover in a balcony and cause the reinforcing bars to rust. When steel bars rust, they expand and push away the thin concrete cover. This will cause cracks, which allow more water to reach the bars. The result is extensive cracking and spalling. Spalling damage is characterized by pieces of concrete falling down from the main body of the balcony. This defect substantially weakens the concrete. If it's not corrected, portions of a balcony can fall.

Slab thickness (inches)	Roof slabs	Floor slabs
2	6x6/6-6 WWM	6x6/6-6 WWM
2½	6x6/4-4 WWM	6x6/6-6 WWM
3	#3 @ 16" o.c.	6x6/4-4
3½	#3 @ 12" o.c.	#3 @ 12" o.c.
4	#3 @ 12" o.c.	#3 @ 15" o.c.
4½	#3 @ 12" o.c.	#3 @ 16" o.c.
5	#4 @ 18" o.c.	#3 @ 13" o.c.
5½	#4 @ 18" o.c.	#3 @ 12" o.c.
6	#4 @ 16" o.c.	#3 @ 12" o.c.
6½	#4 @ 14" o.c.	#4 @ 18" o.c.
7	#4 @ 13" o.c.	#4 @ 16" o.c.
7½	#4 @ 12" o.c.	#4 @ 14" o.c.
8	#4 @ 12" o.c.	#4 @ 14" o.c.

Figure 7-2 Typical temperature reinforcement for one-way slabs

Span (feet)	Thickness (inches)	Reinforcing steel
4	3	#3 @ 7" o.c.
5	3	#3 @ 7" o.c.
6	3	#3 @ 7" o.c.
7	$3^1/_2$	#3 @ $5^1/_2$" o.c.
8	$3^1/_2$	#3 @ $5^1/_2$" o.c.
9	4	#3 @ $4^1/_2$" o.c.
10	$4^1/_2$	#4 @ $7^1/_2$" o.c.
11	5	#4 @ $6^1/_2$" o.c.
12	$5^1/_2$	#4 @ $5^1/_2$" o.c.
13	6	#4 @ 5" o.c.

Figure 7-3 *Typical one-way slab reinforcement*

Here's a checklist of things to consider when you build a cantilevered concrete balcony or exposed elevated concrete slab:

- Is the slab too thin to let you place the steel reinforcing bars properly?

- Is there enough concrete cover over the reinforcing bars?

- Are there aluminum handrail posts close to steel reinforcement bars? This can cause electrolysis that will make the bars corrode.

- Is the concrete so porous that moisture will get in and corrode the reinforcement bars?

- Are the construction joints sealed properly so moisture can't reach the reinforcing?

- Is there too much chloride in the concrete so the reinforcing bars will rust?

- Is there enough waterproof coating on the concrete so the bar won't corrode?

- Is there enough slope for drainage so water won't pond up and penetrate the concrete cover?

Two-Way Flat Slabs

A two-way slab is a concrete slab that spans between two sets of parallel beams, set at 90 degrees to each other. A building with columns whose bays are nearly square may use a two-way flat slab system. A bay is the horizontal space

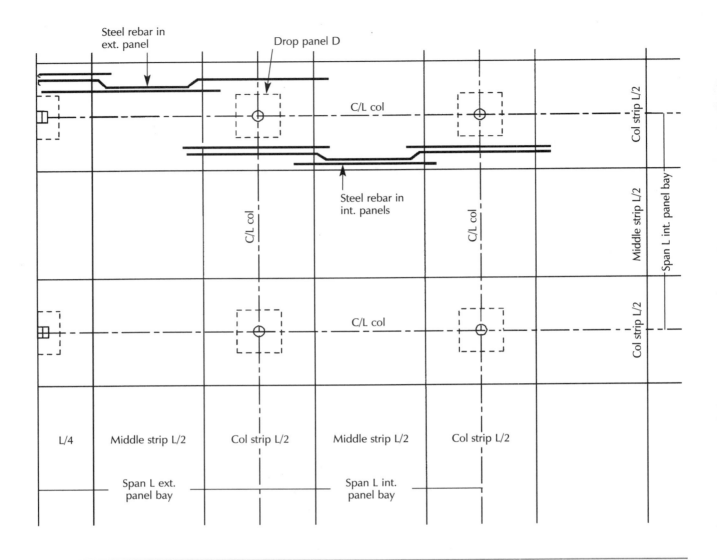

Steel rebar in ext. panel

Drop panel D

C/L col

Col strip L/2

Steel rebar in int. panels

Middle strip L/2

Span L int. panel bay

C/L col

C/L col

Col strip L/2

C/L col

| L/4 | Middle strip L/2 | Col strip L/2 | Middle strip L/2 | Col strip L/2 |

Span L ext. panel bay

Span L int. panel bay

Figure 7-4 Plan of a typical two-way flat slab

between adjacent columns and walls. This kind of slab has load-carrying reinforcement in both directions so you can reduce the thickness of the slab. Some slabs have dropped panels, also called *thickened slabs*, over the columns.

Figure 7-4 shows a typical floor plan with two-way slab construction. You can think of a two-way flat slab as strips of concrete running perpendicular to each other. The slab is divided into panels, and the panels into strips. The strip over the centerline of the columns is called the column strip. The width of the column strip is one-half the span in the opposite direction. The strip between the column strips is called the middle strip. The width of the middle strip is also equal to one-half the span in the opposite direction. The strip alongside the exterior wall or columns is called the exterior strip. It's as wide as one-quarter of the span in the opposite direction. Because each area has different stress conditions, you base the thickness and reinforcement on the loads and the locations of the panels and strips. The abbreviations in Figure 7-4 are:

L = span of slab between supports or columns

$L/2$ = width of column strip in feet

C/L = centerline of columns

Col = column

The load bars at the bottom of a slab are *positive steel*. The load bars at the top of a slab are *negative steel*. Reinforcing steel in two-way slabs is identified as:

- positive steel in a column strip
- negative steel in a column strip
- positive steel in the middle strip
- negative steel in the middle strip
- temperature steel

Figure 7-5 shows typical slab thickness and reinforcement for square bays between 10 and 30 feet square. You may use this table as a guide for preliminary estimating or sizing. The values are based on typical residential floors having a total load of 100 pounds per square foot. You can increase the reinforcement

Spans, range (feet)	Thickness (inches)	Reinforcement, top and bottom
10-13	4	#3 & 6" o.c.
10-14	4½	#4 & 8" o.c.
11-15	5	#4 & 7" o.c.
12-16	5½	#4 & 6" o.c.
13-17	6	#4 & 5½" o.c.
15-20	6½	#4 & 5" o.c.
16-23	7	#5 & 7" o.c.
18-24	7½	#5 & 6½" o.c.
18-25	8	#5 & 6" o.c.
19-27	8½	#6 & 8" o.c.
20-30	9	#6 & 7" o.c.

Figure 7-5 Typical two-way concrete slab reinforcement

spacing slightly for interior spans. Slab thickness also applies to rectangular panels, although the amount of reinforcement is greater in the shorter span and less in the longer span. The maximum ratio between long and short spans is approximately 1.8 to 1.

Metal Pans and Concrete Joists

To reduce the amount of wood forming and shoring you need, use removable metal pans to form a concrete floor or roof. The pans are U-shaped, 20 or 30 inches wide, and 6 to 14 inches deep. See Figure 7-6. The spacing between the pan edges sets the joist width. The sides of an individual metal pan taper inward, so the joists are narrower at the bottom than at the top. The concrete topping over the pans may be 2 to 4 inches thick. This produces a concrete joist that's 8 to 18 inches deep and 4 to 8 inches wide.

Use beams and shoring to support pans. Set the shores at the midspan of the pans. After you place the pans, have ironworkers install the reinforcing bars. Then another crew can cast the concrete in, and over, the pans. Then you can float, screed, and trowel the topping to a smooth finish. When the concrete has almost cured, have carpenters remove the pans and their supports. The concrete joists will then be self-supporting. You can use this system in floor or roof spans from 12 to 30 feet. Figure 7-7 shows a typical concrete pan joist after the steel panels have been removed. The callout '3/4" clear' in the figure means that there should be at least a 3/4-inch concrete cover over the steel bars.

Space supports depending on the size and gauge of the steel pan and the thickness of the concrete you're placing. The heavier the concrete slab and joists are, the closer the shores must be. Check with the pan manufacturer for their recommendations. For example, support for 20- and 30-inch-wide pans would be:

- 2 × 8 soffit planks to support the bottom of the pan
- 4 × 6 stringers to support the soffit plank
- 4 × 4 posts with Ellis clamps to support the stringers. Use the Ellis clamps to adjust the height of posts.

Use stringers to support pans at their ends and at 6- to 7-foot intervals. Space the 4 × 4 posts that support the stringers about 10 feet apart.

Shoring, also called *falsework* or *centering*, supports the formwork and fresh cement. When you strip the forms, leave the permanent post shores undisturbed. You can leave a narrow strip of forms in place directly over each post until the post is removed after the concrete attains its required strength.

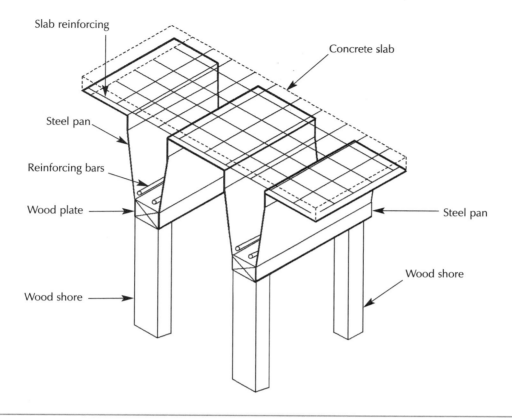

Figure 7-6 *Tapered pan form*

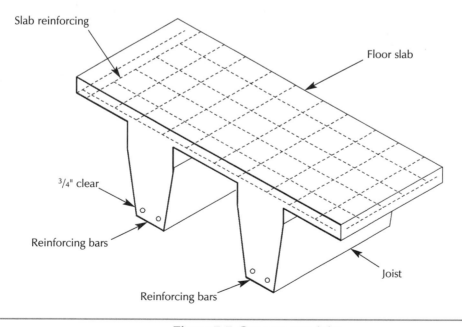

Figure 7-7 *Concrete pan joist*

Reshoring is usually done to make it easier to reuse the formwork by using the strength of the concrete before it has fully cured. You'll remove the form and post shores simultaneously, and replace each individual post immediately with a new post wedged in place to support the concrete. Reshoring is a critical operation and should be planned in advance and approved by an engineer.

There's no single shoring procedure that's standard for all types of buildings. Every job is different. The thickness of a concrete slab and the size of the concrete joists or beams determine the weight that must be supported. The stringers, in turn, are sized by their span, the weight supported, and the carrying capacity. The size of the posts or shores depends on their height. Everything in a shoring system is interrelated.

The usual practice is that the contractor warehouses a stock of standard size stringers, bracing, and shores. The builder plans the shoring layout for each job so that he can use the material in the inventory. Some contractors prefer to rent prefabricated shoring material for a job, including frames, screw jacks, beams, purlins, and connecting hardware. The company that manufactures or rents this material usually prepares the shoring layout for a specific job as part of the rental cost. Most shoring material manufacturers belong to the *Scaffolding, Shoring & Forming Institute*. You can reach them at (216) 241-7333, or on their Web site (www.taol.com/ssfi). It's a good idea to get their brochures on recommended steel frame shoring erection and safety rules.

After the pans are in place, ironworkers install the reinforcing bars. Then, you cast the concrete in, and over, the pans. Float, screed, and trowel the topping to a smooth finish. When the concrete is almost cured, remove the pans and supports. Don't remove the shoring and beam forms too soon. Ideally, you should leave them in place throughout the required curing period. You usually have to strip the forms as soon as possible so you can use them someplace else.

The carrying capacity of a concrete pan joist mainly depends on the width and depth of the joist, the thickness of the slab, the steel reinforcement, and the span of the joist between the beams. For example, for a 20-foot span, you can use a 20-inch-wide by 10-inch-deep pan with a 2-inch-thick slab to carry about 50 psf live load on the floor. A 30-inch-wide by 12-inch-deep pan with a 3-inch-thick slab will carry about 43 psf. Another important factor is the compressive strength of the concrete.

Precast Concrete Joists

Builders have used precast concrete joists for decades to support floor and roof slabs for relatively light loads. But you can't use these joists for heavy or concentrated loads, or where they're subject to heavy impact or vibrations of mounted machinery.

Precast concrete joists are made in a cement products plant and delivered to the job site ready to set in place. The joists come in depths from 6 to 14 inches, and they're shaped like an I-beam. The flange width varies from 2 to 4 inches and the web is about $1^1/_2$ inches. Each joist depth is designed for various strengths depending on the amount of reinforcing steel in the lower flange. For example, an 8-inch concrete joist can contain from 0.22 to 0.88 square inch of reinforcing bars at the bottom, and a $^3/_8$-inch diameter bar at the top with $^1/_4$-inch diameter stirrups. The strength of a precast concrete joist is always calculated and tested by its manufacturer. You can find this information in sales or technical brochures. Most joists are placed 20 to 33 inches on center.

To develop maximum strength, place a 2- or $2^1/_2$-inch-thick slab, or topping, over the top of joists and embed them about $^1/_2$ inch into a concrete slab. Use concrete that has a strength of 3,750 psi for both the joists and the slab.

Forming the Elevated Slab

You can cast a slab supported by precast joists in various types of forms. You can use forms made of plywood supported by wood joists set in hangers. Another type of form you can use is a welded wire mesh with waterproofed paper backing attached. For thin, short concrete slabs, the paper backing serves as the form, and the mesh as the reinforcing. When you place concrete in this type of form, the paper backing sags away from the mesh to the limit of the stitch wires (the parts of the mesh that extend beyond the edge of the paper).

Metal Decking and Concrete Fill

Metal decking installed over steel framing is another economical way to form concrete slabs. You can place corrugated metal sheets between joists or continuously over the top of joists. Because corrugated metal decking can span long distances between steel beams, you'll need little or no shoring. You place and weld the decking to the steel beams. You can also attach steel angles around the perimeter of a deck to control the thickness of a slab. Then place steel reinforcing bars on chairs set on the decking. After the electrical conduits, pipes, and other embedded items are in place and inspected, cast the concrete to a thickness of 4 to 6 inches over the decking. Have concrete finishers complete the job. Figure 7-8 is a detail of a typical concrete slab cast on a metal decking form on a steel frame.

Some types of decking use steel bars or wire welded to the corrugated ribs in the decking. This makes the concrete and deck act together to carry the loads on the slab. It also reduces the amount of added reinforcement. Instead of only supporting the slab, the steel deck gives tensile strength to the concrete. This system is fire rated as 1, 2, 3 or 4 hours. The underside of the steel deck must be

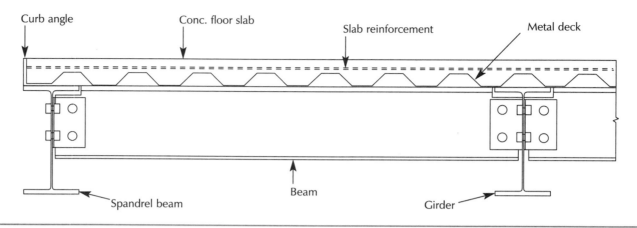

Figure 7-8 Metal decking form on steel frame

protected. You can do this with a suspended plaster ceiling or by applying a fireproofing plaster directly to the bottom of the decking. The thickness of the plaster will determine the fire-resistive rating of the floor.

T-Beams and Spandrel Beams

You can form concrete T-beams, as shown on Figure 7-9, by casting a slab together with closely-spaced rectangular beams. The beams and slab work together as a single structural element. This provides greater strength than casting the beam and slab separately. When the slab terminates at an edge beam, the beam is called a *spandrel beam*. Figure 7-10 shows the recommended bar placement in beams. Figure 7-11 shows various types of spandrel beams and slabs, and their reinforcement.

You'll find more information on beams in Chapter 9.

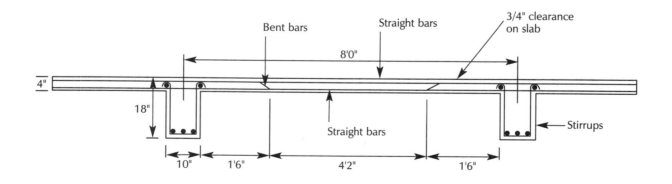

Figure 7-9 Section through typical T-beam

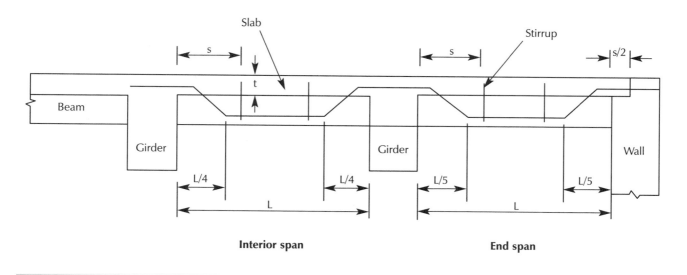

Figure 7-10 *Recommended bar placement in beams*

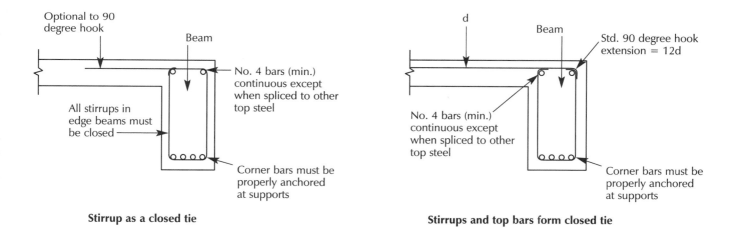

Figure 7-11 *Types of spandrel beam reinforcement*

Precast Slabs and Planks

Precast concrete slabs and planks are manufactured products. The hollow-type units are made with inflated rubber tubes that form circular voids that are half of the cross-sectional area of the unit. (Look back to Figure 5-1 in Chapter 5.) The concrete used in these products is thoroughly vibrated to assure maximum density and strength. Concrete may be either regular weight or lightweight. It's cured in heated kilns to make it cure faster and more completely. Figure 7-12 shows some typical sizes, spans, and loads for regular weight concrete planks. Check Figure 7-13 for typical supports of precast concrete planks.

These units are usually delivered to a job site by truck and set in final position by crane. After the slabs or planks have been placed side by side, they're aligned and the keyways in the sides are filled with grout. The price of a unit is normally quoted to include installation and caulking of the joints. In earthquake prone areas, the general contractor usually installs concrete topping.

For long span concrete floor slabs, you can use units that are rectangular, inverted U-channels, T-beams, double T-beams, or hollow-core planks. You can get prefabricated hollow-core planks 4 to 6 inches thick which are designed to span as far as 45 feet. You can cover the planks with a 2- to 3-inch-thick concrete topping. The advantages of these precast units are:

- all-weather construction
- no forming or shoring
- shorter construction time
- longer spans
- low weight
- ease of erection from truck to building

Figure 7-14 shows a hollow-core precast concrete plank. You can use these for floor and roof slabs. In areas of seismic activity, you need the concrete topping to tie the deck together.

Size (depth/width, inches)	Spans (feet)	Load (psf)
6 x 24	15-25	43
8 x 24	20-33	57
10 x 20	28-40	61
10 x 24	28-40	72
12 x 24	35-50	79

Figure 7-12 *Typical sizes, spans and carrying capacities of precast planks*

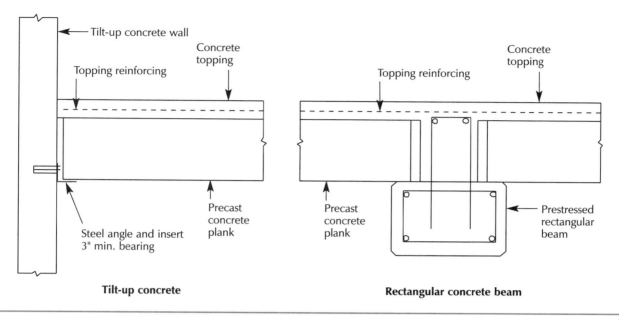

Figure 7-13 *Precast concrete plank supports*

An example of a precast hollow core plank is the *Spancrete Plank*. These range in thickness from 4 to 10 inches and span up to 40 feet. You can use them with no topping or with a $2^1/_2$-inch-thick concrete topping. These planks are pre-engineered and are popular for precast concrete buildings. You can install them over steel or concrete beams, concrete or block walls, or cast-in-place walls. For example, a 4-inch-thick plank will carry 162 psf on a 10-foot span, a 6-inch-thick plank will carry 56 psf, an 8-inch plank can carry 52 psf, and a 10-inch-thick plank can carry 58 psf. The strength of the planks varies according to the size and number of wires in the plank.

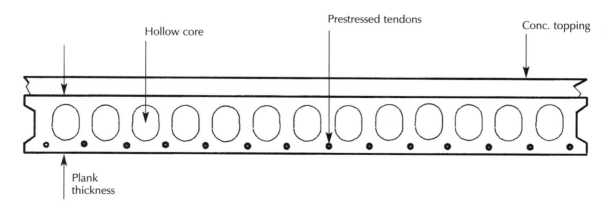

Figure 7-14 *Section through a typical hollow-core concrete plank*

Omnia Widespan Panels

An Omnia Widespan panel is a specially-reinforced precast concrete panel, only 2 inches thick, that you can install on bearing walls or beams. Using these panels, you don't need any wood or steel forming. The panels serve as forms on which you can cast 2 to 8 inches of concrete topping. Because a system of wire truss reinforcement embedded in the precast panels projects into the topping, the final slab acts as a single structural unit.

Chapter 8

Columns and Walls

Columns support roof and floor loads, and transfer the total weight of a building to its foundation. They're also called piers, pedestals, posts, pillars, or supports. Columns are classified as long or short columns. The length of a long column is between 10 and 11 times its width. The length of a short column is more than 3 times, and less than 10 or 11 times, its width.

A pier is shorter than a short column, and a pedestal is shorter than a pier. The height of a pier or a pedestal is less than 3 times its narrowest width. The term "post" is usually used with a wood column.

A column's height is limited by the ratio of its height to its smallest width or diameter — the *slenderness ratio*. This ratio also affects the column's load-carrying capacity. Ideally, you should try to use a short column with an unsupported height less than 10 times its width. In most cases, you're free to increase the column's width, but you can't increase its height. For example, make a 12-foot-high column no less than 144 / 10, or 14.4 inches wide, rounded up to 15 inches. For a 14-foot column, use 16.8 inches wide (168 / 10), rounded to 17 inches.

The allowable load on a short column with minimum reinforcement is equal to its cross-sectional area times 25 percent of the ultimate strength of the concrete, or:

$$P = A \times .25f'c$$

If you have more than minimum reinforcement in a concrete column, the steel will carry some of the load. The carrying capacity of a reinforced concrete column is:

$$P = A[1 + (n-1)\,p]\,fc$$

where:

P = total load capacity of the column, in lbs.

A = cross-sectional area of the column, in sq. in.

n = ratio of the modulus of elasticity of steel, Es, to that of concrete, Ec. Value of n varies from 15 to 10 in 2000 to 3000 psi concrete.

p = percent of steel cross-sectional area to concrete area

fc = allowable unit working stress of concrete, in psi

If you're restricted to a maximum column width that would exceed the h / d ratio of 10, then the capacity of the column is reduced by this formula:

$$P'/P = 1.33 - h/(120\,R)$$

where:

P' = allowable load on a long column, in lbs.

P = allowable load on a short column, or A × .25 fc, in lbs.

h = unsupported length of column, in inches.

R = radius of gyration of the column cross-section. For a square column, R = 0.288675d. For a round column, it's d / 4.

Here's a sample calculation for a plain, or unreinforced, pier or pedestal:

$$P/A = fc = 0.25f'c$$

where:

P = load on pier or pedestal, in lbs.

A = cross-sectional area of pier or pedestal, in sq. in.

fc = allowable unit working stress of concrete, in psi

$f'c$ = 28-day compressive strength, or ultimate strength, of concrete, in psi

Columns are loaded either axially or eccentrically. An axially-loaded column supports a load applied close to the center of the column. An eccentrically-loaded column supports an off-center load. It must have more reinforcement than an axially-loaded column to resist that off-center load. Figure 8-1 shows the load-carrying capacity of columns of various sizes and reinforcement.

Column size, in.	Load on bars, kips min. to max.	Load on 2000 psi concrete, kips	Load on 3000 psi concrete, kips
12 x 12	23 to 92	52	78
14 x 14	33 to 125	71	106
16 x 16	41 to 164	92	138
18 x 18	52 to 207	117	175
20 x 20	64 to 256	144	216

Note:
Column reinforcing bars grade $f's = 20,000$ psi
Minimum area of reinforcement $As = .008 \times Ag$
Maximum area of reinforcement $As = .032 \times Ag$
Where Ag = cross-sectional area of rectangular column

Formulas:
Total capacity $\quad$ P (kips) = (0.18 f'c Ag + 0.8 fs Ag) / 1000
Load on bars $\quad\quad$ P (kips) = 0.80 fs Ag / 1000
Load on concrete $\quad$ P (kips) = 0.18 f'c Ag / 1000

Example:
Assume load (P) is 300 kips.
Try 18-inch square column
$\quad$ Ag = 18 x 18 = 324 sq. in.
$\quad$ Select concrete strength f'c = 3000 psi
$\quad$ Load on concrete $\quad$ P = 0.18 x 3000 x 324 / 1000 =174.96 or 175 kips
$\quad$ Remainder of load carried by bars = 125 kips
Try 8 #8 bars (area of #8 bar is 0.79 sq. in.)
$\quad$ As = 8 x 0.79 = 6.32 sq. in.
$\quad$ Load carried by bars $\quad$ P = 0.8 x 20,000 x 6.32 / 1000 =101.12 kips
$\quad$ Not enough: Try 10 #8 bars As = 7.9 sq. in. $\quad$ P = 0.8 x 20,000 x 7.9 / 1000 = 126.4 kips
OK. Use 10 #8 bars in 18-inch square column.

Figure 8-1 Load-carrying capacity of columns and reinforcement

Column Reinforcement

There's a minimum and maximum amount of steel reinforcement you can use in a column. The total cross-sectional area of vertical steel bars in a column should be at least 1 percent, and not more than 8 percent, of the column's area. For example, a 12-inch-square column has an area of 144 square inches. The total cross-sectional area of the vertical steel bars in it must be more than 1.44 square inches, but less than 11.5 square inches.

Be especially careful placing reinforcement in columns. Columns are more vulnerable to high stresses than beams or slabs. Most structural failures in a building after a severe earthquake occur in its columns. Figure 8-2 shows various types of columns and reinforcement.

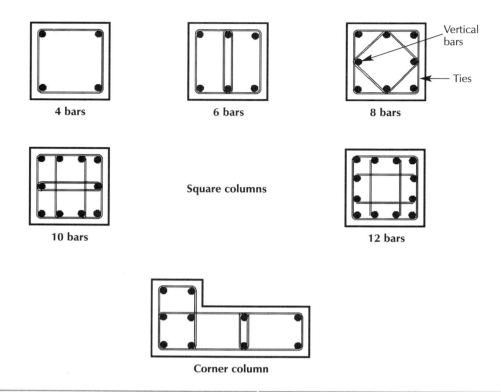

Figure 8-2 *Typical column reinforcement*

In multistory buildings, the columns at the top floors carry the lightest loads and the columns at ground level carry the heaviest loads. You need to add more reinforcing steel in the ground level columns so they can be the same size as the other columns in the building.

Column reinforcing steel runs both vertically and horizontally. Vertical bars carry part of the loads on a building.

Most columns fail because the vertical reinforcing bars buckle. You need to use steel ties or hoops to encircle the vertical bars to keep them from buckling outward and breaking, or spalling, the concrete. The size, number, spacing, and hooking of column ties are very important. Space column ties no more than:

- 16 times the diameter of the vertical bars in the column

- $1/2$ of the smallest side of the column in a rectangular column

- 48 times the diameter of the ties in the column

For example, if a column has four #8 vertical bars, the nominal diameter is 1 inch. Therefore, the ties should be spaced a maximum of 16d, or 16 inches apart. If a column has six #6 vertical bars, the nominal diameter is $3/4$ inch, so space the ties a maximum of $16 \times 3/4$, or 12 inches apart. Ties in a 12- $\times$ 16-inch rectangular column should be no more than 6 inches apart ($1/2 \times 12$). Space ties $3/8$ inch in diameter no more than 18 inches apart ($3/8 \times 48$).

Rectangular Columns

A rectangular column can be square or rectangular. Square columns may have 4, 6, 8, 10, 12, 14, 16, 18, or 20 vertical bars, depending on the size of the bars, the column, and the applied load. You hold the vertical reinforcing bars in place with ties or hoops. Usually the ties are made of $^3/_8$- or $^1/_2$-inch-diameter plain steel bars. Normally you space ties 8 to 12 inches apart.

Circular Columns

Most builders prefer circular columns because they're easier to form. They usually use fiber cylinders, 6 to 48 inches in diameter, as column forms. These cylinders are made of spirally-laminated plies of fiber with a wax coating on the exterior and interior surfaces. The wax coating makes them weather- and moisture-proof.

Reinforcing bars in round columns are normally in a circular pattern rather than a rectangle. Ties are spiral-wound No. 2 gauge wires spaced about $2^1/_4$ inches apart. You also use circular loops as ties. Typical reinforcement for a 16-inch-diameter column is six #6 bars. An 18-inch round column usually contains eight #6 bars. The concrete that covers the ties should be at least $1^1/_2$ inches thick.

In special cases, like when you don't have much space for a conventional concrete column, it's better to use a concrete-encased steel column. This may be a steel H-beam encased in a concrete column. This type of column is also called a *composite column.* The steel core H-beam or pipe will carry most of the load. If you use a steel pipe, fill the interior with concrete. The concrete cover also provides the fire protection.

Designing Columns

The basic formula for designing an *axial-loaded* column is:

$$fc = .25 f'c$$

where:

fc = the compressive stress on a column, in psi

$f'c$ = 28-day compressive strength, or ultimate strength, of concrete, in psi

That means the allowable compressive strength is $^1/_4$ of the ultimate strength of the concrete.

As an example, the stress on an 8-inch-square plain column with a 50,000-pound load is:

$$fc = 50,000 \ / \ (8 \times 8)$$

$$= 781 \ psi$$

So the concrete in the column must have an ultimate compressive strength of at least 3,500 psi (781 $\times$ 4 = 3124 psi).

The factors you use in formulas for designing axial-loaded columns are:

P' = total safe axial load on a long column

P = total safe axial load on a short column

h = unsupported length of the column

d = minimum width of the column

A = cross-sectional area of a column

R = minimum radius of gyration of a column. This is the square root of I / A.

I = moment of inertia of the column. The moment of inertia, I, of a square column is equal to d^4 / 12. The value I of a rectangular column is equal to b $\times$ d^3 / 12, and I for a round column is equal to 0.049087 $\times$ d^4, or πr^4 / 4, where r is the radius of the column.

Take the following steps in designing a concrete column:

1. Figure the load (P) the column is to carry.
2. Find the height of the column, or clear distance between floor slabs, beams or other lateral supports (h).
3. Decide on the concrete strength you'll use (fc = .25f'c).
4. Try a column size that is $^1/_{10}$ of the height of column.
5. Use longitudinal bars of at least 1 percent of cross-sectional area of column (p).
6. Divide the load (P) by the cross-sectional area (A) to find the unit compressive stress (fc).
7. If fc is more than .25 f'c, increase the column size or add more reinforcement.
8. If you add more reinforcement, check load capacity of a reinforced concrete column by using this formula:

$$P = A \ (1 + (n - 1) \ p) \ fc$$

where:

n = 15 for 1500 to 2200 psi concrete, 12 for 2201 to 2900 psi concrete, and 10 for over 2901 psi concrete

p = ratio of the cross-sectional area of the reinforcing bars to the cross-sectional area of the column

Before you start concrete work, be sure you prepare plans and specifications for concrete columns that clearly show these items:

1. Size, number, location, and grade of reinforcement bars.

2. Splicing of reinforcing bars (butted or lapped).

3. Minimum spacing of bars and bundles of bars.

4. Offsets between column faces in multistory buildings with different-size columns.

5. Bending of vertical reinforcing bars that tie a column to the connecting beams. Plans should include a bending schedule for each type of reinforcement.

6. Spirals in round columns. Spirals are continuous wires formed into a helix that hold the vertical bars in place. Plans should note the spacing of the wires, outside diameter of the spiral, and how the spirals in a long column are connected to each other. See American Concrete Institute publication ACI 315-17.

7. Bundled ties. Use bundled ties, or preassembled cages, in rectangular columns to hold the vertical bars in place.

Reinforcing steel fabricators usually prepare Column Placing Drawings that show column schedules, supplemented by sketches and tables. These indicate the bending and location of all bars and ties. The schedule shows each tier of column reinforcement from the foundation to the roof, including:

1. Overall size.

2. Number, size and length of straight and bent bars in each column.

3. Bending schedule showing each part of the bar which is indicated by a letter shown on the sketch.

4. Mark number for each type of straight or bent bar, tie, and spiral.

Walls

The design for an exterior concrete wall depends on the load and the required fire-resistance rating. The National Fire Protection Association (NFPA) and the Underwriters Laboratories publish standard ratings of concrete walls.

Wall thickness (inches)	Size and spacing of bars in each direction
6	#4 @ 13" o.c. center wall
8	#4 @ 18" o.c. each face
10	#4 @ 15" o.c. each face
12	#4 @ 12" o.c. each face

Figure 8-3 *Typical concrete wall reinforcement*

The fire-resistance rating is the time it takes a fire on one side of a wall to ignite combustible material on the other side of the wall. Here are some of their standards:

■ Fire-resistance rating of a concrete wall varies from 1 to 4 hours, depending on the thickness of the wall.

■ Fire-resistance period of a solid concrete wall is 4 hours for a $6^{1}/_{2}$-inch-thick wall; 3 hours for a 6-inch wall, 2 hours for a 5-inch wall, and 1 hour for a $3^{1}/_{2}$-inch wall.

Usually you place a single layer of horizontal and vertical bars at the center of a 6-inch-thick wall as the steel reinforcement in the wall. Use two layers of bars, one at each face, in thicker walls. As a rule of thumb, the minimum reinforcement for any concrete wall should be #4 bars placed 12 inches on center, both horizontally and vertically. Figure 8-3 shows the minimum reinforcement for concrete walls 6 to 12 inches thick.

Precast Walls

Using precast concrete construction is an economical way to build walls. Concrete wall panels are usually cast on-site but may be precast off-site at a fabricating plant. When you make them on-site, you usually use a building floor slab as the casting platform. If the floor has many obstructions, such as trenches or pedestals, build a temporary concrete slab nearby for casting panels. Figures 8-4 and 8-5 show how to tie precast concrete walls together.

Here's how tilt-up construction generally proceeds:

■ Operating engineers rough grade the building site.

■ Laborers place and fine-grade a bed of crushed rock.

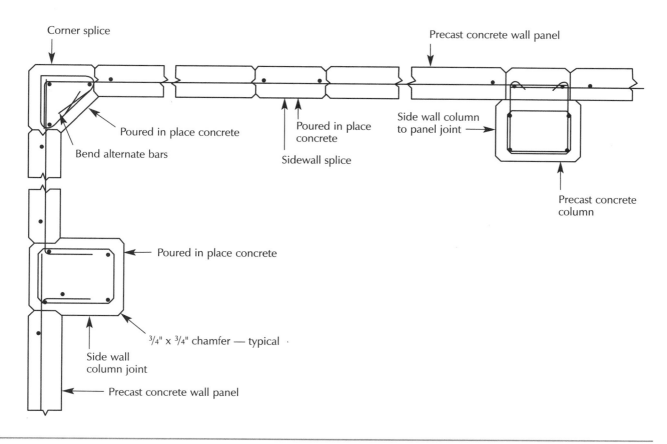

Corner splice

Poured in place concrete

Bend alternate bars

Poured in place concrete

Sidewall splice

Precast concrete wall panel

Side wall column to panel joint

Precast concrete column

Poured in place concrete

3/4" x 3/4" chamfer — typical

Side wall column joint

Precast concrete wall panel

Figure 8-4 Precast concrete joints

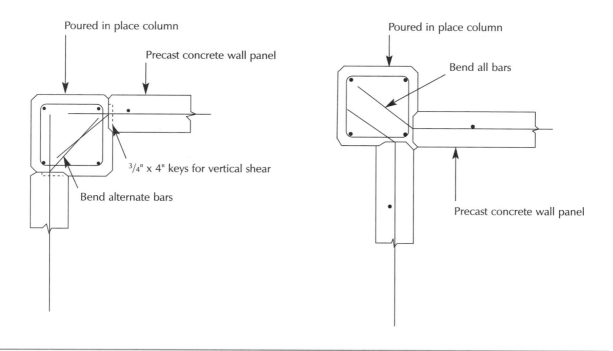

Poured in place column

Precast concrete wall panel

3/4" x 4" keys for vertical shear

Bend alternate bars

Poured in place column

Bend all bars

Precast concrete wall panel

Figure 8-5 Precast concrete corner details

- Carpenters install wood edge forms.

- Ironworkers place reinforcing bars, dowels, and anchors.

- Laborers place and screed concrete floor slab.

- Concrete finishers float and finish concrete slab.

- Slab cures.

- Laborers make all joints in the slab smooth with mortar to provide a good casting platform for wall panels. There should be no indentations or projections.

- Carpenters set wood forms on the casting platform, and spray the forms and slab with an antibonding agent. *This is very important.* You don't want to discover that the floor slab and wall you're about to lift into place are one solid unit. It's happened.

- Ironworkers place the reinforcing steel, anchor bolts, inserts, and all necessary hardware in their proper positions in the panels. Inserts include those needed for lifting and bracing the panels.

- Laborers place concrete into the forms.

- Finishers float and screed to level the surface of the concrete to the form top. They trowel the surface to a smooth finish. A texture finish can be applied at this time.

- Finishers spray a curing agent on the surface of the freshly-cast concrete to reduce the rate of drying.

- When the concrete has fully cured, use a mobile crane to raise the panels and place them on the footings or floor slab in their final upright position. Working room for the crane is important. Be sure you have enough space to maneuver the crane around to lift and set each concrete panel.

- Laborers brace the panels plumb to the floor slab.

- Ironworkers place column reinforcing steel.

- Carpenters install column forms.

- The mobile crane operator transports concrete from the transit-mix truck by bucket to the wall columns and places concrete into the column forms.

- Column concrete cures.

- Carpenter strips column forms.

- Laborers backfill over foundations and place concrete in the floor strip (a 3- to 4-foot-wide strip of floor slab adjacent to the wall that's left open until the precast panels have been set in place and braced). That finishes the floor and embeds the dowels extending from the panels.

Tilt-up construction works well for warehouses, manufacturing facilities, and simple box-like structures. Its advantages are:

■ Formwork costs less than cast-in-place walls.

■ You can cast one slab over another that's cured. This cuts down the space you need at a job site. Use this method when there's not enough floor space to cast all panels at one time.

■ Entire construction takes less time.

■ It requires less quality control.

■ You can use more common and/or unskilled labor.

■ You can cast foundations while the precast panels are curing. When you use a remote casting platform, you can place concrete in the foundations later.

The disadvantages of tilt-up construction are:

■ Precasts weigh more.

■ Precasts require more temperature reinforcement.

■ More design is required for pickup points and extra reinforcement.

■ Precasts require more reinforcement and hardware for pickup stresses.

■ Floor slabs must be free of trenches, slopes, cracks, and other obstructions.

■ A more expensive, high-capacity crane is required.

■ Careful planning is required to provide access and maneuvering room for the mobile crane.

Retaining Walls

For most soils, friction will hold back stationary soil at an angle of about 37 degrees to the horizontal. At a greater slope, you need retaining walls to keep soil from sliding, or to resist the pressure of earth, sand, or other filling.

Retaining walls are always working. Soil pressure is constant. But pressure increases when soil gets wet or a vehicle passes by a wall. The condition of the subgrade under a wall's footing will affect how well the wall resists sliding or overturning. You have to keep water from penetrating the soil adjacent to a footing. If the subgrade becomes soft, the wall may rotate and collapse.

Here are some of the ways retaining walls can fail:

■ Wall slides horizontally.

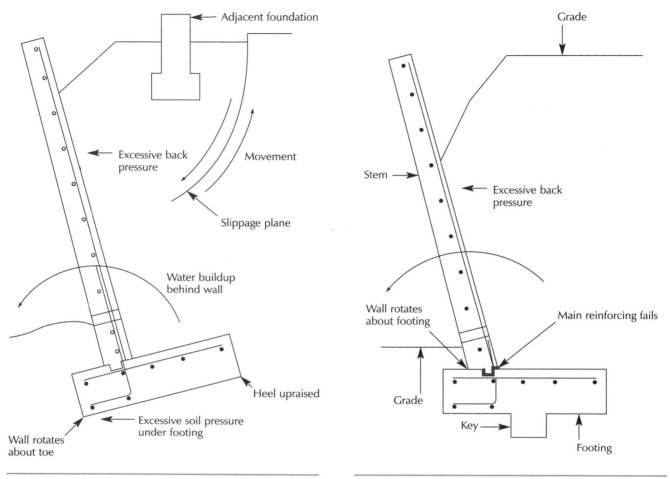

Figure 8-6 *Overturning failure of a retaining wall*

Figure 8-7 *Structure failure of a retaining wall*

■ Wall rotates about the toe of the footing.

■ Wall fails structurally.

■ Water penetrating into the soil increases the pressure behind the wall.

■ Unexpected surcharge (an additional vertical load) increases the pressure on the wall. A surcharge may be caused by a sloped backfill, vehicular traffic, or the weight of an adjacent building.

Figure 8-6 shows a wall failure, where the wall overturns either from excessive back pressure from an adjacent building or from water buildup behind the wall. Figure 8-7 shows how a bond failure in the reinforcement can cause the stem of a wall to rotate.

Types of Retaining Walls

Generally speaking, there are two types of retaining walls: gravity and cantilevered. The gravity wall depends on the weight of the concrete and the friction at the bottom of the concrete to resist the pressure of the backfill. Figures 8-8 and 8-9 show two types of gravity retaining walls. Note that there are weep holes through the walls to keep water from accumulating and adding pressure on the walls. To help you size a typical gravity retaining wall, Figure 8-10 lists approximate dimensions and volumes of concrete for different gravity wall heights.

Cantilever retaining walls have footings to provide more contact area with the soil, and are shaped to resist overturning and sliding. They're either T- or L-shaped. Both shapes have a stem, or wall, and a footing. The thickness of the stem increases from top to bottom. The top is usually 6 to 8 inches thick. Cantilever retaining walls are more common because they require less concrete, though more reinforcing steel.

To picture the footing of a retaining wall, think of the wall as facing away from the embankment it holds. The part of the footing under the retained earth is called the *heel*. The opposite part of the footing is called the *toe*. The friction of the footing and the pressure of the soil against the toe of the footing should keep a wall from sliding. Some footings also have a vertical protrusion under the footing, called a *key*. The key helps keep a wall from sliding. If the embankment soil at a

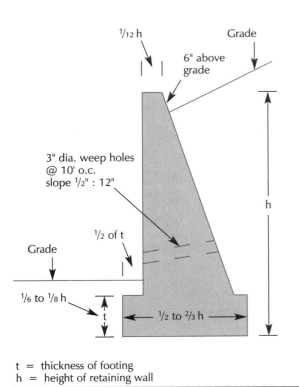

t = thickness of footing
h = height of retaining wall

Figure 8-8 *Gravity retaining wall with footing*

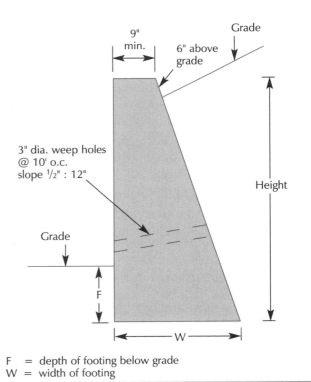

F = depth of footing below grade
W = width of footing

Figure 8-9 *Gravity retaining wall without footing*

Height (feet)	Depth (inches)	Width (feet)	Volume (C.Y./lineal foot)
3.5	9	2.06	0.182
4.0	10	2.25	0.222
4.5	11	2.44	0.266
5.0	12	2.63	0.313
5.5	13	2.81	0.363
6.0	14	3.00	0.417
6.5	15	3.19	0.474
7.0	16	3.38	0.535
7.5	17	3.56	0.600
8.0	18	3.75	0.667

Figure 8-10 *Typical gravity retaining walls*

wall may become moist, you add a key to the bottom of the footing to make the wall stronger. Deadman anchors are used to prevent overturning of a retaining wall when you can't excavate an adequate footing.

Select the size of the heel and toe to keep a wall from overturning. You should design retaining walls to provide a safety factor against overturning of at least 150 percent. If there isn't enough stabilizing weight, increase the length of the heel. Also, if the resultant vector comes too close to the toe, increase the length of the toe. (See the section on using graphics to design retaining walls.)

Also, you should design the depth of the footing and key of a retaining wall to keep it from sliding. Sliding of the retaining footing should be resisted by the toe's vertical face. When this isn't adequate, a key at the bottom of the footing increases the sliding resistance.

Backfill and surcharges can make a wall slide. You can compensate for a highway loading surcharge by adding 2 more feet to the height of soil behind a retaining wall when you calculate the loads on the wall. Figure 8-11 shows the main components of a typical cantilever retaining wall with a surcharge.

Pressure on a Retaining Wall

All retaining wall design has to consider the amount of pressure on the wall. Normal dry earth puts about 30 pounds per square foot for every foot of wall below the grade at the top of the wall. A stiff cohesive clay produces less than 30 psf per foot of wall below the surface of the clay. When soil is very fluid, or almost approaching mud, it has a pressure the same as water, or 62.4 psf per foot of depth. To prevent a buildup of hydrostatic pressure on a wall, put weep holes at

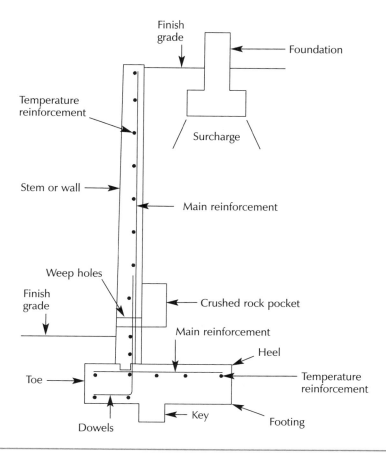

Figure 8-11 *Components of a cantilever retaining wall with surcharge of adjacent building*

the bottom of the wall. When a wall must retain a possible landslide, some codes require you to use a pressure of 125 psf per foot of wall below the surface when you design the wall. This means the wall should be able to resist 40 percent more pressure than is actually behind it.

Steel Reinforcing in a Retaining Wall

The main steel reinforcement in a retaining wall is at the back side of the stem and at the bottom of the footing. You need at least 3 inches of concrete cover when you cast concrete against soil. The horizontal reinforcing in the wall and the footing should help keep the concrete from cracking as it shrinks and expands due to changes in temperature. It also helps hold vertical bars in place. Figures 8-12 and 8-13 show typical reinforcement for retaining walls without a surcharge. Figure 8-14 shows a cantilever retaining wall without surcharge. Figures 8-15 and 8-16 show typical reinforcement for retaining walls with a surcharge.

You can use Figures 8-10 through 8-16 to guide you in your preliminary sizing for retaining walls of various heights and loading conditions. All horizontal bars in these walls are #4 rebar, placed 13 inches on center.

Height (feet)	Wall thickness (inches)	X bars	Z bars
4	8	#4 @ 13" o.c.	—
5	8	#4 @ 13" o.c.	—
6	8	#4 @ 13" o.c.	—
7	8	#4 @ 13" o.c.	—
8	8	#4 @ 11" o.c.	#4 @ 13" o.c.
9	8	#5 @ 11" o.c.	#4 @ 13" o.c.
10	10	#5 @ 12" o.c.	#5 @ 14" o.c.
11	10	#6 @ 12" o.c.	#5 @ 14" o.c.
12	10	#6 @ 9" o.c.	#5 @ 14" o.c.
13	12	#6 @ 9" o.c.	#5 @ 12" o.c.

Figure 8-12 *Typical retaining wall without surcharge (wall reinforcement)*

Height (feet)	Depth (inches)	Base (feet-inches)	Toe (inches)	Rebars (y bars)	Rebars (t bars)
4	12	2'3"	6	#4 @ 13"	3 #4
5	12	2'8"	6	#4 @ 13"	4 #4
6	12	3'0"	6	#4 @ 13"	4 #4
7	12	3'6"	9	#4 @ 13"	5 #4
8	12	4'0"	12	#4 @ 11"	5 #4
9	12	4'4"	12	#5 @ 11"	6 #4
10	16	4'9"	12	#5 @ 12"	6 #5
11	16	5'3"	12	#5 @ 12"	6 #5
12	16	5'9"	12	#5 @ 9"	7 #5
13	16	6'3"	15	#5 @ 9"	7 #5

Figure 8-13 *Typical retaining wall footing without surcharge (footing reinforcement)*

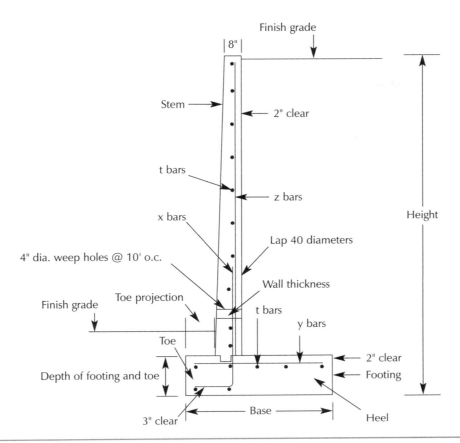

Figure 8-14 Concrete cantilever retaining wall without surcharge

Height (feet)	Wall thickness (inches)	X bars	Z bars
4	8	#4 @ 13" o.c.	—
5	8	#4 @ 11" o.c.	#4 @ 11" o.c.
6	8	#5 @ 11" o.c.	#4 @ 11" o.c.
7	10	#5 @ 11" o.c.	#5 @ 11" o.c.
8	10	#6 @ 12" o.c.	#5 @ 12" o.c.
9	10	#6 @ 9" o.c.	#5 @ 9" o.c.
10	12	#6 @ 9" o.c.	#5 @ 9" o.c.
11	16	#6 @ 11" o.c.	#6 @ 11" o.c.
12	16	#6 @ 14" o.c.	#6 @ 14" o.c.
13	16	#6 @ 13" o.c.	#6 @ 13" o.c.

Figure 8-15 Typical retaining wall with surcharge (wall reinforcement)

Height (feet)	Depth (inches)	Base (feet-inches)	Toe (inches)	Rebars (y bars)	Rebars (t bars)
4	12	3'6"	9	#4 @ 13"	5 #4
5	12	4'0"	12	#4 @ 11"	5 #4
6	12	4'4"	12	#5 @ 11"	6 #4
7	16	4'9"	12	#5 @ 11"	6 #5
8	16	5'3"	12	#5 @ 11"	6 #5
9	16	5'9"	12	#5 @ 11"	7 #7
10	16	6'3"	15	#5 @ 9"	7 #5
11	20	7'0"	18	#6 @ 12"	7 #6
12	20	7'4"	18	#6 @ 12"	8 #6
13	20	7'8"	24	#6 @ 10"	8 #6

Figure 8-16 Typical retaining wall with surcharge (footing reinforcement)

Using Graphics to Design a Retaining Wall

Figure 8-17 shows how to use graphics to design a typical 10-foot retaining wall. Here's how to figure the vectors of forces shown on Figure 8-17:

1. W1 is the weight of the wall figured at 150 pcf ($10 \times .83 \times 150 = 1245$ lbs.). Draw an arrow, or force vector, to scale and place it at the center of gravity of the wall. If you use a scale of 1 inch equals 1000 pounds, the vector W1 would be 1.245 inches long.

2. W2 is the weight of footing ($4.75 \times 1.33 \times 150 = 948$ lbs.). Place this vector at the center of gravity of the footing.

3. W3 is the weight of the soil over the heel of the footing using 100 pcf ($10 \times 2.92 \times 100 = 2920$ lbs.) Place this vector at the center of gravity of the block of earth over the heel of the footing.

4. Draw the three vertical vectors, W1, W2, and W3 to scale in a row beside your sketch.

5. Add or scale these vectors to find the total vertical forces. The sum of vertical forces is about 5113 pounds.

6. To locate the position of the total vertical force, multiply W1, W2, and W3 by their respective distances from the toe of the footing:

$[(1245 \times 1.42) + (948 \times 2.375) + (2920 \times 3.29)] / 5113 = 13626 / 5113 = 2.66$ ft.

7. F is the total of all forces from the retained earth pressing against the wall. It is based on 30 psf per foot of depth below finish grade. For this 10-foot-high wall, the horizontal force at the top of the wall is zero, 1 foot below it is 30 psf, and at the top of the footing, the force is 300 psf. These forces are indicated by the triangle drawn at the side of the wall. The total horizontal force is the area of the triangle (10 × 300 × .5 = 1500 lbs). Place this vector at the centroid of the triangle, which is one-third the height of the wall (10 / 3 = 3.33 ft).

8. Multiply the 1500-pound force by its vertical distance above the toe of the footing (1500 × 4.66 = 6990 ft-lbs). This gives you the total overturning moment.

9. The ratio of stabilizing moment to overturning moment is 13626 / 6990 = 1.95. A factor of safety of at least 1.5 is required, so the dimensions are acceptable.

10. Draw the F vector at the top of the three vertical vectors and draw a diagonal which closes the force diagram.

11. The diagonal line is the resultant of forces on the wall, in scale and direction. It passes through the footing at the first third point. As the center third of the footing is between 1.58 and 3.16 feet from the toe, the retaining wall is stable.

The advantage of the vector system is that it's simple and fast. You can scale length of vectors and the distances from the toe of the footing to see whether the retaining wall is stable, without lots of calculations.

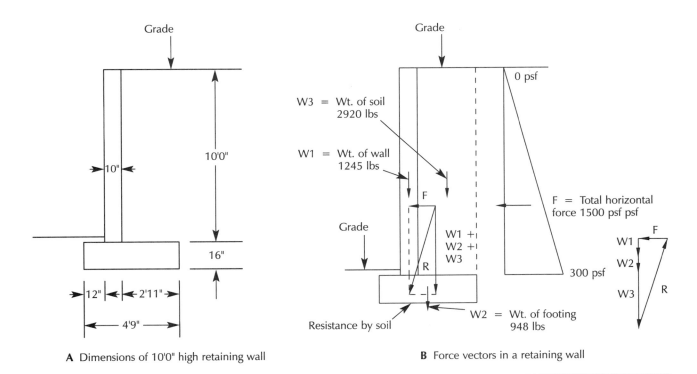

A Dimensions of 10'0" high retaining wall   **B** Force vectors in a retaining wall

Figure 8-17 Using graphics to design a retaining wall

Basement Walls

In most subterranean buildings, the exterior walls serve as retaining walls. That means these walls must resist the inward push of the earth, as well as support the building. Figure 8-18 shows a single level subterranean wall. Figure 8-19 shows a two-level basement. When subterranean walls span from basement to first floor, make sure both the basement floor slab and the first floor slab are in place and cured before you place any backfill material at the site.

Waterproofing Basement Walls

Although basements with concrete walls and floors may be structurally sound, they're defective if they're flooded during a rainy season. This happens if a building site has a high water table, or if the adjoining ground is porous. There are two basic basement waterproofing systems you can use to avoid this problem. The drained (or pumped) system and the undrained (or barge) system.

The Drained System — Let's look first at a drained or pumped system. This system uses a conventional wall and floor slab with a drainage bed, sump, and pump to keep the water away from basement walls as much as possible. You can use this system in tight, dense soil when the rate of water inflow is less than the pump's capacity.

In a drained system, it's better to use reinforced concrete walls than reinforced concrete block walls because they're more impervious. Cast-in-place concrete walls are usually 8 inches thick and reinforced with horizontal and vertical bars. Floor slabs are usually 4 inches thick and reinforced with welded wire mesh. A gravel bed of 4 to 6 inches is usually enough. Build a concrete sump pit with several drain holes in all four sides. Install a pump in the pit. Make a gravel collecting pocket near the footing on all walls exposed to subterranean water. Also, install a gravel-filled slot, 2 inches deep by 4 inches wide, through the footing every 6 to 10 feet. Make the slot the same length as the width of the footing. This allows water that accumulates in the outside pocket to reach the sump pump.

You don't need a drain tile in a properly-designed gravel underdrain. However, if you use a 4- or 6-inch tile pipe, install it around the exterior basement walls so that the bottom of the drainage pipe is higher than the footing. You want it to intercept subterranean water before the water can pass under the footing and into the basement area. If you use short lengths of pipe, make $1/2$- to $3/8$-inch open joints. Cover the top half of the joints with bitumen-saturated paper or felt to keep fine particles from entering the pipe and clogging it. You can also use continuous lengths of perforated pipe laid with the perforations facing downward.

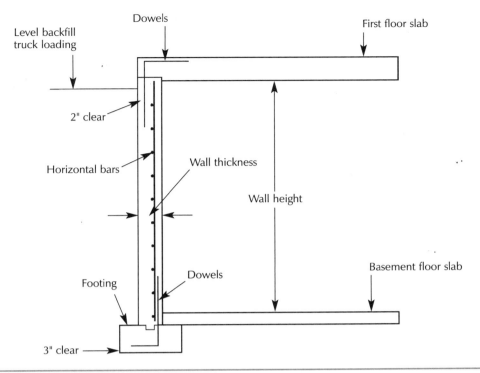

Figure 8-18 *Concrete basement wall*

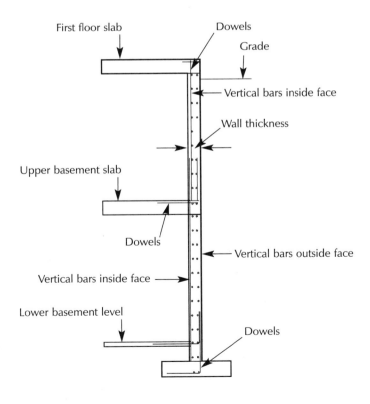

Figure 8-19 *Two-level basement wall*

The Undrained System — The other waterproofing system is called the undrained, or barge system. Use this method when the soil is very porous and a normal pump can't match the rate of inflow. You've got to design the basement of an undrained system to be watertight, like the hull of a barge. Install a complete waterproof envelope around the walls and floor slab. Make all joints in the slab, between the slab and wall, and around pipe penetrations watertight with rubber water stops, caulking, and nonshrinking concrete grout. Be sure to prevent any punctures during construction and backfill.

Undrained basements on building sites with a high water table can become buoyant. You may have to increase the thickness of the walls and slab to get additional weight to offset the upward pressure of the water under the slab. A 6-inch-thick reinforced slab is usually adequate to resist a water pressure of 5 feet of water. The pressure at a depth of 5 feet is 5×62.4, or 312 psf.

Industrial Concrete Walls and Columns

In many industrial facilities, concrete walls and columns come in contact with chlorides, sulfates, and other corrosive compounds. You need to use concrete that can resist such substances. Also, you may have to cover the concrete with a polymer or epoxy protective coating. You may want to use fiber-reinforced plastic or epoxy-coated steel reinforcing bars. You can get information on the durability of concrete in hostile environments from:

US Army Corps of Engineers

Department of the Army
Portland District Corps of Engineers
ACASS Center
P.O. Box 2946
Portland, OR 97208-2946
503-808-4590
Fax 503-808-4596
www.usace.army.mil

Directorate Military Programs
Construction & Engineering Division
20 Massachusetts Ave. NW
Washington, DC 20034-1000
202-761-0336
www.hq.usace.army.mil

Portland Cement Association

5420 Orchard Road
Skokie, IL 60077
847-966-6200
Fax 847-966-9781
www.portcement.org

American Concrete Institute

P.O. Box 9094
Farminghill, MI 48333
248-848-3700
www.aci-int.net

Concrete Reinforcing Steel Institute

933 No. Plum Grove Road
Schaumburg, IL 60173-4758
847-517-1200
Fax 847-517-1206
www.crsi.org

Chapter 9

Beams and Girders

There are generally three types of concrete beams: simple beams (beams with noncontinuous ends), continuous or fixed-end beams, and cantilever beams. Concrete beams may be reinforced with conventional mild steel at the bottom and top. Also, reinforced concrete beams may be pre-stressed or post-stressed. The shape of concrete beams may be rectangular, or with an integral slab such as an upright or an inverted T-beam.

A *simple concrete beam* tends to crack or break at the bottom of the beam. This is because the load on the beam makes the top of the beam compress and the lower part of the beam stretch. Concrete can resist the compression, but you need steel reinforcement to resist the stretching at the bottom of the beam.

A *continuous beam* will sag between its supports and bend up over the supports. Tensile stress on the beam is at the bottom of the beam at midspan, just like a simple beam, but also at the top of the beam at the supports. You need reinforcement at the bottom, at midspan, and the top of a continuous beam over the supports.

In *cantilever beams,* tensile stress is at the top of the beam while the lower part of the beam is in compression. So you need reinforcement at the top of a cantilever beam.

Another type of beam is the *T-beam*. It's similar to a concrete joist, only much larger. The flange at the top of a rectangular T-beam acts like a slab spanning between beams, except it's an integral part of the beam. It helps resist part of the compressive stresses in the beam. Figure 9-1 shows the main parts of a T-beam: the stem or web, and the flange. The bottom figure illustrates the tendency for cracks from vertical loads on the slab. The top figure shows the reinforcing bars needed to prevent the cracking.

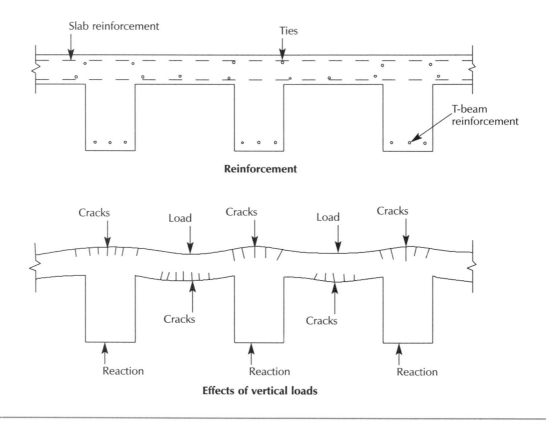

Figure 9-1 *Effects of vertical loads*

For symmetrical T-beams, the maximum width of a flange is $1/4$ the span length of the beams, or 8 times the thickness of the slab, or $1/2$ the distance between adjacent beams. For one-sided T-beams, the maximum width of span is 6 times the slab thickness, or $1/2$ the distance between adjacent T-beams. For isolated T-beams, the maximum width of a flange is 4 times the width of the beams.

Girders are generally larger than beams. They usually support the ends of a series of beams. Columns, pilasters, or walls usually support girders. The principles for designing girders are the same as those for beams. A girder may also be simple, continuous or cantilever.

Pre- and Post-tensioned Beams

Pre- or post-tensioning a beam increases its strength. To make a pre-tensioned concrete member, stretch steel tendons (also called *wires* or *rods*) over a casting bed within the forms. Place the steel parallel to the casting bed. Pour concrete in the forms, embedding the stretched tendons. After the concrete has fully cured, cut the tendons at each end of the member. Fabricators use this method to mass produce pre-stressed concrete beams.

Contractors can make post-tensioned concrete members in the field. Just place hollow tubes containing the unstretched tendons or specially wrapped wires in the forms before you pour the concrete. Position the tensioning strands in a parabolic form. After the concrete has been placed and has cured, stretch the cables and anchor them to the ends of the beam. Use this method to make long-span and unrepeated members, and special items such as beams and girders that are more than 75 feet long.

For circular concrete tanks, silos, and large-diameter pipes, use a special self-propelled machine that winds stressed wires continuously around a cylindrical structure to make post-tensioned concrete. Protect the wires with pneumatically-applied concrete.

You can use uncoated stress-relieved strands for pre-tensioned bonded designs. These strands are usually $^1/_4$ to $^1/_2$ inch in diameter, with an ultimate strength of 9,000 to 36,000 pounds. The galvanized strands for post-tensioned concrete are from 0.600 to $1^{11}/_{16}$ inches in diameter, with ultimate strength of 46,000 to 352,000 pounds.

Wire sizes are usually given in decimal parts of an inch because the increments are so small. Bars are usually given in fractions. The decimal 0.6 inch = $^5/_8$ inch. The fraction $1^{11}/_{16}$ inch = 1.6875 inch. A number classification of a deformed steel bar stands for the number of $^1/_8$-inch increments in the diameter of the bar. For example, a No. 5 bar is $^5/_8$ inch in diameter and a No. 7 bar is $^7/_8$ inch in diameter.

Here are some requirements for pre-stressing steel that's bonded to concrete:

■ The maximum diameter of a single-wire strand is 0.2 inch.

■ The maximum diameter of a 7-wire strand is $^3/_8$ inch.

■ Center-to-center spacing of wire should be at least 3 times the wire or strand diameter.

■ Make sure there's a clear space between the wires that's at least $1^1/_2$ times the maximum size of the coarse aggregate. See Figure 9-2.

■ Make sure there's a clear space from the edge of the bar to the surface of the concrete that's at least $1^1/_2$ inches, or one bar diameter, whichever is greater. For beams where the tension in the pre-stressed steel is end anchored, the above requirements don't apply, except that the distance from the side of the beam to the surface of the nearest bar should be $1^1/_2$ times the maximum size of the coarse aggregate.

■ Use stirrups with maximum spacing less than $^3/_4$ of the beam depth.

■ Cross-sectional area of stirrups should not be less than 0.80 percent of the beam's cross-sectional area.

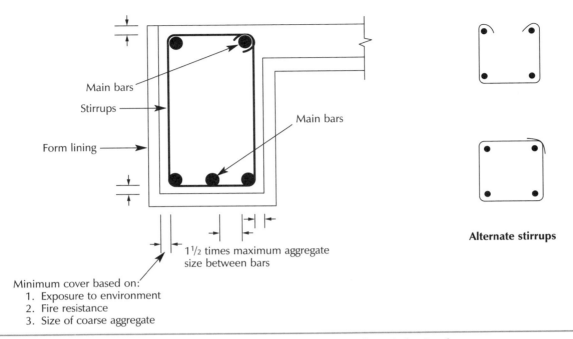

Figure 9-2 *Minimum clearances for reinforcing bars*

Steel Reinforcement in Beams

Here are some rules on anchoring steel reinforcement in beams:

■ For simple and cantilever beams, extend at least $1/3$ of the steel reinforcement 6 inches into supports.

■ For continuous beams, extend at least $1/4$ of the steel reinforcement 6 inches into supports. Extend every reinforcing bar at least 12 bar diameters beyond the point where it's not needed.

Under certain conditions, you should place steel bars in the upper part of a beam. This will help the beam resist stress and may often allow you to use a smaller beam. Figure 9-3 shows the cross-section of two beams, one with tensile reinforcement only and one with both tensile and compression reinforcement (or *doubly reinforced*). Here are some symbols you'll commonly find on drawings like this:

h = total depth of the beam from top to bottom

d = effective depth of the beam, the distance from the top of the beam to the center of the tensile reinforcement

kd = distance from the top of the beam to its neutral axis

jd = distance between the centroid of the compression area to the center of the tensile reinforcement

fc/n = ratio of the tensile stress to the ratio of Es/Ec

fc = the maximum compressive stress in the beam

n = ratio of Es/Ec

As = areas of tensile steel

A's = areas of compressive steel

nA's is the area of concrete in tension (under the elastic theory)

When a beam carries vertical loads, it will tend to bend due to a bending moment, or M (see Chapter 11). This causes compressive stresses in the upper part of the beam and tensile stresses in the lower part.

The compressive stress varies from a maximum fc, at the outer fibers of the beam, to zero at the neutral axis. The resultant of the compressive stress is called C, which acts in the centroid of the stress triangle.

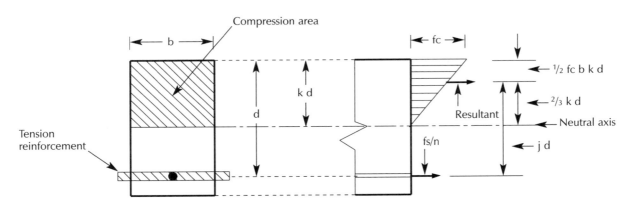

Beam with tension reinforcement

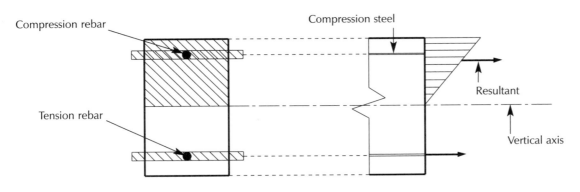

Beam with tension and compression reinforcement

Figure 9-3 Concrete beam force diagram

As concrete is assumed weak in tension, the tensile reinforcement (called T) carries the entire tensile stress. Forces C and T are equal and opposite. The distance between C and T creates a couple called the resisting moment, RM, of the beam. RM and M are equal and opposite.

The designer selects the beam size (b and d), strength of concrete (f'c), strength and amount of reinforcement (As) to make the resisting moment of the beam equal to the applied moment without exceeding the strength of the concrete (fc), or steel (fs).

Here are some general rules for the minimum amount of reinforcement you need in beams and girders:

- Minimum tension reinforcement should be more than 0.005 times the effective depth of the beam times the width of the beam.

- Minimum compression reinforcement should be more than 0.005 times the effective depth of the beam times the width of the beam at the outer end of the beam, built integrally with supports.

Use the following rules for placing bars in building construction:

- Keep the space between bars at least 1 bar diameter, or 1 inch.

- Keep at least $1^1/_2$ times the maximum size aggregate, or 1 inch clear space, between parallel bars, as shown in Figure 9-2.

- On beams exposed to weather or soil, maintain at least $1^1/_2$ inches of cover over main bars or stirrups near the surface.

- For highway construction, space bars $1^1/_2$ times the bar diameter and make concrete cover at least 2 inches thick.

Stirrups

All beams should also have web reinforcement, or stirrups. They are $^3/_8$- or $^1/_2$-inch plain bars bent in a U-shape, or as closed loops, spaced at intervals. The spacing of stirrups shouldn't exceed d/2, where d is the effective depth of the beam. The minimum spacing of the stirrups shouldn't be less than b/2, where b is the width of the beam. Stirrups hold longitudinal bars in place and keep a beam from separating from its supports. Figure 9-4 shows diagonal bars. On the figures, L is the span of the beam.

You should design a beam for shear stress as well as bending stress. If the shear stress is more than the shear strength (fv) of the concrete, stirrups are required. Bent longitudinal bars can also be used with the stirrups. To find the shear stress in a beam, divide the total shear by the cross-sectional area of the beam at that point, or v = V/bd. Chapter 11 describes how to find shear, V.

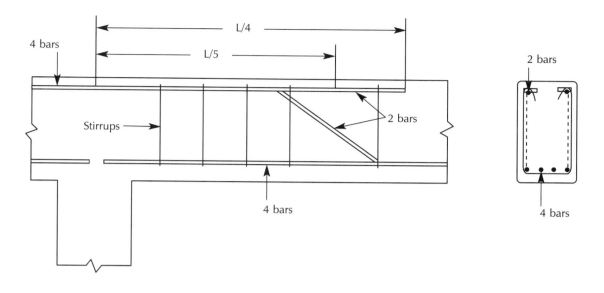

Figure 9-4 *Shear bars*

The required number and spacing of stirrups depend on the shape and area of the shear diagram for which stirrups are required. Use these steps:

1. Select the type of stirrup (vertical or inclined), the size Av (total area of stirrup legs) and allowable steel stress (fv).

2. Calculate v'b / B Av of the large and small ends of the shear diagram and points of discontinuity.

3. Calculate S, length of the base of the shear diagram.

4. Determine the number and spacing of the stirrups.

Drawings and Specifications for Concrete Beams

Drawings should contain beam schedules that list each beam's mark number and size, plus the number and size of straight and bent bars. Each beam is identified on the placement plans that the engineer, reinforcing steel fabricator, or rebar installer prepares. The beam mark number is assigned by the steel detailer, who may be an independent contractor, or employed by the steel fabricator.

Plans should also have notes describing the bending, size and spacing of stirrups or stirrup ties. The notes should describe the location of top bars, and where to use truss bars.

It's often easier to design the size and number of reinforcement bars in a beam than to install them. When a cluster of bars are too close together, concrete can't flow around them. Air pockets can develop under the bars, or the bars may be too close to the surface of the beam, causing spalling.

Concrete design plans should show the following information:

■ Type and location of all splices and mechanical devices.

■ Use of standard ACI bends and hooks.

■ Where straight and bent bars are alternated, show the spacing as follows:

#4 @ 10" alt str & bent (meaning space bars 10 inches on center for adjacent bars).

■ When concrete cover is specified, indicate whether it's to the main bars, stirrups, or ties.

■ When beams intersect other beams or columns at the same plane, indicate the clearances required for the bars to pass each other.

Designing Beams and Girders

Critical points differ in different beams. Heavily-loaded short beams are more likely to crack. Narrow beams over long spans are more likely to buckle. When you design a beam, it's a good idea to use the tables prepared by the American Concrete Institute instead of complex formulas. You'll save time, and you're less likely to make mistakes. Figure 9-5 is a typical table for sizing beams. Most tables for sizing concrete beams incorporate formulas that show the relationship of beam

f'c / n	Fc (psi)	K	k	j	P
2000 / 15	900	157	0.403	0.866	0.0091
2500 / 12	1125	196	0.403	0.866	0.00113
3000 / 10	1350	236	0.403	0.866	0.0136
3750 / 8	1700	298	0.403	0.866	0.0172
5000 / 6	2350	393	0.403	0.866	0.0229
Note: (fs = 20,000 psi)					

Figure 9-5 *Table for sizing beams*

size, concrete and steel strengths, and amount and location of steel reinforcement. Commonly-used concrete strengths (f'c) are 2000, 2500, 3000, 3750, and 5000 psi. Usual allowable compressive stress in concrete is based on fv = 0.45f'c.

Here are other formulas for balance design:

$$K = fc/2\ k\ j$$

$$d^2 = M\ /\ K\ b$$

$$M = d^2 K b$$

$$A = fs\ /\ 12000\ (average\ j\ value)\ for\ use\ in\ As = M\ /\ ad$$

$$J = 1 - k\ /\ 3$$

$$R - \tfrac{1}{2}\ fc\ k\ j = p\ fs\ j$$

$$Fc = 0.45\ f'c$$

You should keep in mind that you're designing against potential failure. When designing a beam, ask yourself the following questions:

- How many stirrups and diagonal bars do I need near the end of the beam to keep it from cracking?

- Do I have enough reinforcement in the beam to keep it from cracking or being crushed by a heavy load? Remember, you need reinforcement at the bottom of a simple beam, and near the top, at the supports of a continuous beam.

- Do I have enough lateral support for the beam to keep it from buckling?

- Is the beam wide enough to keep it from buckling under a heavy load?

- Are the bars and the concrete full bonded together?

- Have I lapped the bars properly so they act like a continuous bar? Figure 9-6 shows the right and wrong ways to lap bars. See Figure 9-7 for minimum lap for bars.

Here are some suggestions on how to design a concrete beam:

1. Find the approximate dead and live load on the beam.

2. Find the clear distance (L) between supports. This may be the face of the columns, walls, or girders.

3. Check and see if there are any architectural restrictions, such as maximum depth of beam for minimum headroom.

4. Pick an approximate size for the beam from a table.

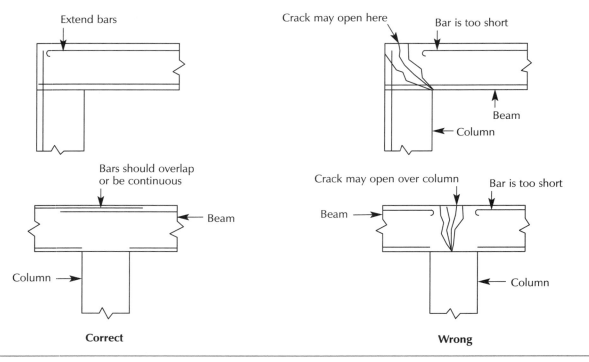

Figure 9-6 *Design errors*

5. Calculate the dead load of the beam by multiplying its cross-sectional area by 150 pounds per cubic foot.

6. Calculate the dead weight of the slab supported by the beam.

7. Calculate the reaction (R) at each end of the beam. This is the shear load (V).

8. Calculate the maximum bending moment (M) of the beam based on the dead load plus live load and type of beam. It may be a simple beam, fixed end beam, or a continuous beam.

9. Check the unit shear stress (fv) in the beam. If it's over the allowable shear stress, increase the depth of the beam.

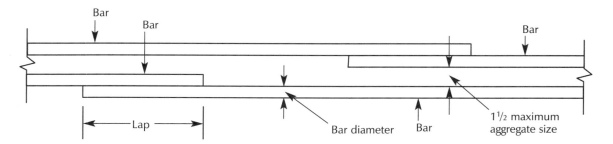

Minimum lap for tension bars is 25 diameters
Minimum lap for vertical column bars is 20 diameters

Figure 9-7 *Lapped bar splices*

10. Check the maximum compressive stress (fc) in the beam due to the bending moment (M). If it's over the allowable stress; increase the width or the depth of beam, or both. Check the maximum tensile stress (ft) in the beam due to the bending moment. If it's over the allowable stress, increase the depth of beam or add tensile reinforcement.

11. Recalculate beam with revised dimensions.

Using Software to Design Beams

A quick and easy way to design beams, as well as other concrete structures, is to use computer software. The Concrete Reinforcing Steel Institute (CRSI) and Integrated Structural Software, Inc. (ISS) are two organizations that have developed software to help you design concrete:

- beams

- columns

- cantilever retaining walls

- solid one-way slabs and joists

- square footings and pile caps

- two-way solid flat slabs, solid flat plates and joists (waffles)

These programs can help you design using any concrete strength or grade of reinforcement steel you choose. Two modules of ISS's concrete engineering software — for basic concrete elements like beams and columns — are included on the CD inside the back cover of this book. Just enter your requirements and the program will calculate the size of the member you need, or check if the member you've selected is appropriate under standard engineering practice, and under normal conditions. For more information about engineering software, contact ISS on their Web site at www.robot-structures.com, or CRSI at www.crsi.org.

Formulas Used in Concrete Design

All prepared tables are developed from formulas that determine the stresses occurring in beams under load. Some of these formulas are shown in Figure 9-8. The symbols used in these formulas are shown in the building codes, ACI manuals, and other references for concrete design and construction. These symbols will also appear in reports on concrete mix design and test results. Figure 9-9 shows these symbols and what they mean.

Design methods have changed several times in recent years. Previously, most concrete design was based on working stress that involves limiting the stress on a concrete member to a fraction of its ultimate strength. The current method uses the ultimate strength of the concrete, but also increases the different kinds of

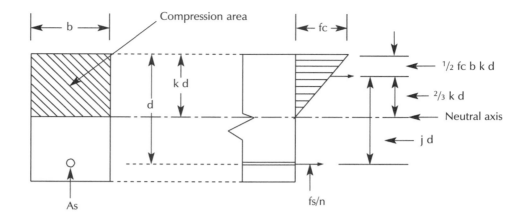

Formulas:

$d = \sqrt{M / kd}$

$As = p\,b\,d = M / f\,s\,j\,d$ of reinforcing steel

$fs = 2\,M / k\,j\,b\,d^2 = M / As\,j\,d$

Figure 9-8 Stresses in a reinforced concrete beam

fc	compressive stress in concrete, psi	*A's*	effective area of compressive steel reinforcement, psi
ft	tensile stress in reinforcement, psi	*Σo*	sum of perimeters of group of bars, sq. inch
fv	tensile stress in web reinforcement, psi	*s*	spacing of vertical stirrups, inch
fa	axial bearing stress, psi	*b*	width of rectangular beam, inch
V	total shear, lbs.	*b'*	width of web of T-beam, inch
v	unit shear stress, psi	*d*	depth from compressive surface to tensile reinforcing steel center, inch
vc	unit shear stress permitted on concrete of web, psi	*d'*	depth from compressive surface to compressive reinforcing steel center, inch
v'	v − vc, psi	*M*	bending moment or moment of resistance, in-lbs.
u	bond, psi		
f'c	ultimate concrete strength, psi	*k*	ratio of depth to neutral axis to depth, d
Es	modulus of elasticity of steel reinforcement, psi	*j*	ratio between distance from center of compression to center of tension and depth, d
Ec	modulus of elasticity of concrete, psi		
n	Es / Ec	*p*	ratio of effective area of tensile reinforcement couple to effective area of concrete, As / bd
R	M / bd², lbs.		
fs	tensile stress in reinforcement, psi	*p'*	ratio of effective area of compressive reinforcement couple to effective area of concrete, A's / b d
As	effective area of tensile steel reinforcement, psi		

Figure 9-9 Notations for reinforced concrete

loads by a factor. The advantage to this method is that the loads are adjusted by their effect on a structure. For example, earthquake and wind loads are short duration, while live loads are long duration and affect the structure differently. A building can absorb short duration loads better than long duration loads.

Embedding Pipes in Beams

It's common practice for electrical, mechanical, and plumbing contractors to install their conduits, pipes, and ducts into the beam forms before the concrete is cast. There are some precautions you should take if this is done on your job. If you embed pipes containing liquid, gas or vapor in concrete, place the pipes between the upper and lower reinforcement. Also, test the pipes to $1^1/2$ times the design pressure, or a minimum pressure of 150 psi for 4 hours before placing concrete. Maximum pressure and temperature is 200 psi and 150 degrees F. Don't fill pipes with liquid, gas, or vapor until the concrete has thoroughly set. Don't allow any liquid, gas, or vapor in the pipe that's detrimental to concrete.

Concrete beams and girders are the most important structural elements in a building. If a girder fails, it affects the connecting beams, joists, and floor slabs. So, it's important to design, detail, and check all plans carefully. When reinforcing steel fabricators submit their placement drawings for approval, check them carefully. They make mistakes too. This is the last chance for you to pick up any design errors. Building inspectors will inspect the forming and reinforcement before they let you pour any concrete, but they're not engineers. And once you pour the concrete, it's too late to correct errors inexpensively.

Chapter 10

Miscellaneous

Concrete Structures

There are many concrete structures we see every day that are not part of a building — including concrete pavements, curbs, gutters, sidewalks, driveways, catch basins, and concrete crib walls. To build most of these items you need to use expansion and contraction joints, anchor bolts, steel ledgers, and other elements, just as you would in a building.

Pavements, Curbs and Gutters

A concrete roadway may be from 5 to 8 inches thick, depending on the traffic it carries. The local road department usually decides the thickness. Most residential or local concrete roadways have a 6-inch-thick aggregate base under the pavement. They're 40 feet wide from curb face to curb face. Secondary highways are 64 feet wide, while major highways have two 42-foot-wide lanes with a dividing strip or median separating the lanes.

You can use a laser-controlled machine to place large concrete pavements. The machine disperses concrete with an auger, then vibrates and consolidates the concrete. Finally, it screeds and finishes with an 8- to 12-foot-wide screed mounted on a 20-foot-long telescopic arm. The machine can finish about 100 square feet of pavement in one pass.

Curbs

The main function of a concrete curb along a roadway is to control drainage and to make an obstacle to help keep vehicles from inadvertently leaving the roadway. You can build a concrete curb as a stand-alone item, or integral to a

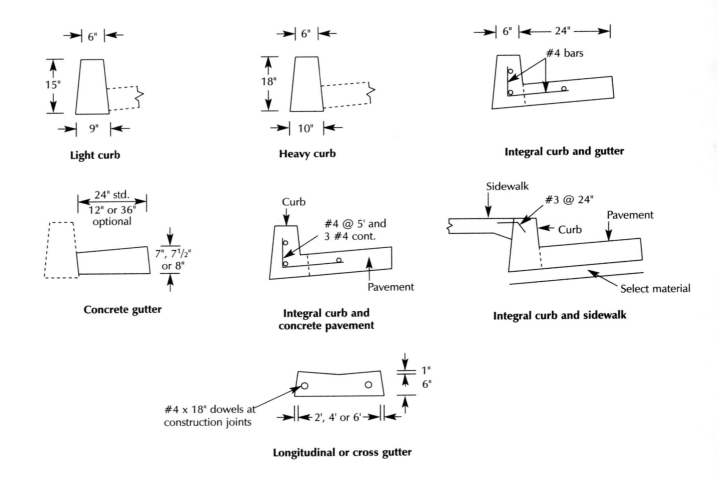

Figure 10-1 *Curbs and gutters*

concrete gutter. In some cases, you'll want to include a curb with a sidewalk or concrete pavement. For example, on commercial properties, sidewalks are often built right up to the curb. But landscaped parkways between the curb and the sidewalk have an advantage because there aren't any obstructions like signposts, fire hydrants, and meters in the way of pedestrian traffic.

A curb is usually 6 inches wide at the top and 9 to 10 inches wide at the bottom. Most curbs are 6 inches high except those that have to be higher to control the flow of rainwater. Light-duty curbs are 15 inches high and heavy-duty traffic curbs are 18 inches high. See Figure 10-1.

The exposed vertical side of a curb is called the *face*. The line where the face meets the top of the pavement or gutter is called the *flow line*. The slope or grade of a flow line is very important in maintaining a clean gutter. An ideal slope is $1/4$

percent. You should finish the lower 4 inches of a face with a steel trowel to make a smooth finish so debris will flow easily along the curb. Also, round off the top edges of a curb to a $^3/_4$-inch radius to help keep the concrete from spalling.

Gutters

You can cast concrete gutters beside a curb, or along with the curb. Usually it costs less to cast it along with the curb. The formwork will cost less and you only place and finish the concrete once, instead of twice.

Most gutters are 12 or 24 inches wide and 6 to 7 inches thick, and the top surface slopes toward the flow line. You may reinforce a curb and gutter unit with #4 bars. Use three longitudinal bars, one near the top of the curb and one at the outer edge of the gutter. Put the third longitudinal bar where you use crossbars (short #3 steel bars) to tie the sidewalk and curb together. Place crossbars 5 feet on center.

A *curb return* is the curved section of a curb that occurs at the intersection of roads. Set expansion joints at the beginning and end of the curve and 60 feet on center between curb returns.

Install contraction joints in a gutter at 25-foot intervals. Make scoring joints $^1/_4$ inch deep in a curb wherever there are contraction joints in the gutter. *Scoring joints* are the transverse lines you see on most concrete sidewalks. They're also called *weakened joints* or *control joints*.

Cross gutters continue the flow between curbs and gutters, and across street intersections. They're usually 6 to 7 inches thick and 24 inches wide. Make the flow line in the center continuous with adjacent flow lines. Connect separate pours of a cross gutter with dowels made of two #4 bars, 18 inches long.

Sidewalks and Driveways

Concrete sidewalks are usually 4 inches thick, except where you dowel them to a curb. In this case, use #3 dowels, 12 inches long at 24 inches on center. Make the sidewalk 5 inches thick over the dowels. It's a good idea to cast a sidewalk on a select base of sand or decomposed granite. Residential sidewalks are usually 4 feet to 6 feet wide and 6 inches off the property line. Use full sidewalks on commercial properties.

A sidewalk should slope toward the street at a $2^1/_2$ percent grade. The back edge of a sidewalk is usually very near a property line, as most sidewalks are on public property. When you cast a sidewalk with a street catch basin, place the sidewalk and the top of the catch basin as a single unit.

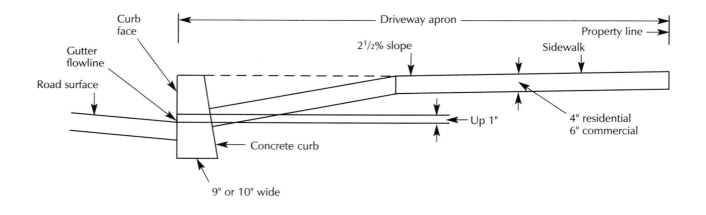

Figure 10-2 *Driveway section*

Driveways

Technically, a driveway is a part of the road right-of-way line and the curb, or the area between the right-of-way line and the pavement (if there's no curb) where motor vehicles enter or leave the public street onto private property. The traveled portion of a driveway and sidewalk is called a *driveway apron*. See Figures 10-2 and 10-3.

A driveway should be at least 10 feet wide, not including the side slopes and return. The maximum width of a residential or commercial driveway is 20 feet, if the property is less than 100 feet wide. If the property is more than 100 feet wide, the maximum width is 30 feet, or 20 percent of the width of the property, whichever is greater. The minimum thickness of a concrete driveway on residential property is 4 inches, and on commercial property it's 6 inches. Make a rough broom finish on all concrete driveways. Scoring lines on driveways should correspond with the scoring lines in adjacent sidewalks.

Catch Basins

Water collected in a catch basin should flow across a sloping floor, then through a connection pipe to a storm drain, usually near the center of a street. The connection pipe may be placed in any position around the walls of the catch basin as long as they point in the right direction and slope at the proper grade. Steel angles that are embedded in the walls support the grating.

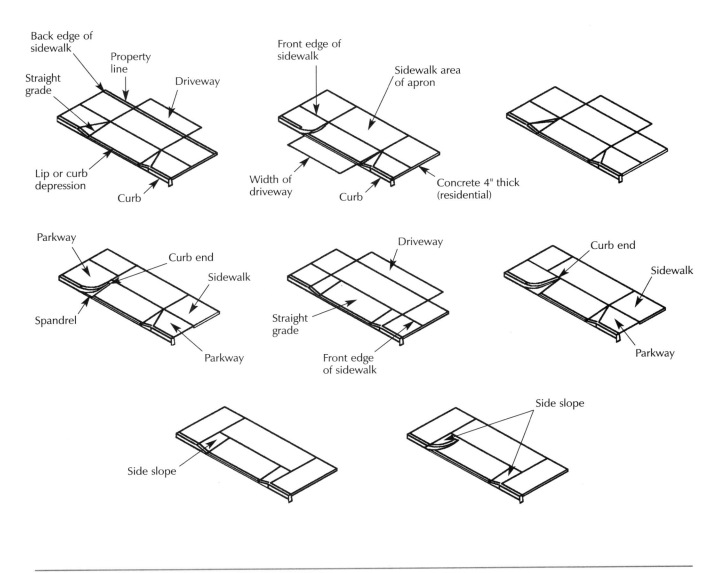

Figure 10-3 *Driveway parts*

A catch basin has 6-inch-thick walls, a 6-inch-thick sloping floor slab, and various embedded steel items. It's about 3^1/$_2$ feet deep. See Figure 10-4. You should give the floor of a basin a steel trowel finish. When a catch basin is contiguous with a new sidewalk, use the same type of concrete for the top of the basin and the sidewalk. In this case, you can leave out the dowels between the wall and the top slab. Give the tops of catch basin walls a smooth finish.

There are also small-area catch basins used in industrial facilities. These are built with 6-inch-thick walls and floor, and reinforced with #6 bars in both directions. A heavy steel grating and frame covers the top to resist traffic passing over the basin.

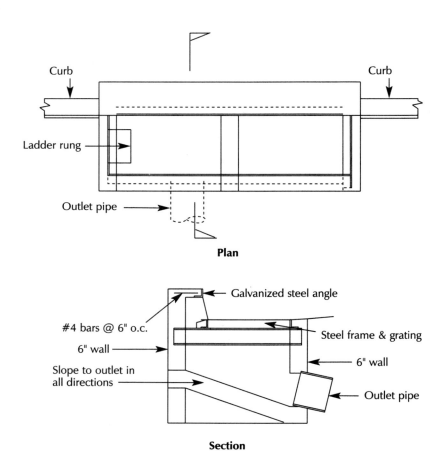

Plan

Section

Figure 10-4 *Catch basin*

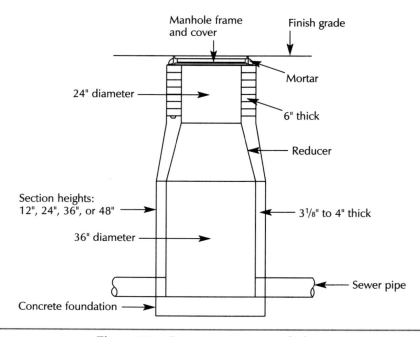

Figure 10-5 *Precast concrete manhole*

Manholes

Precast concrete manholes have generally replaced the old brick manholes. They're quickly set in place by a crane, and adjustable to most depths. There are three major parts to a precast concrete manhole: the lower 3-foot-diameter cylinder, the conical transition part, and the upper 2-foot-diameter rings. See Figure 10-5.

The lower section is made of units reinforced with $1/4$-inch bars placed 4 inches on center. The walls of the section are $3^1/8$ to 4 inches thick. The lower sections can be 12, 24, 36, or 48 inches long. Each section locks into the lower section in a tongue-and-groove manner.

The conical transition unit reduces the inside diameter from 36 inches to 24 inches. Above the transition portion, you have a choice of a number of precast concrete rings to reach the road elevation — $2^1/2$, 6, or 8 inches in height. Just make sure the total height of the rings isn't more than 18 inches. Set the cast iron manhole frame in mortar at finish grade level.

Manholes are covered with cast iron manhole covers that rest on the upper rings. A poured-in-place concrete footing supports the entire manhole.

Crib Walls

Use crib walls instead of retaining walls for very high embankments. They can be made of timber or precast concrete. We're only concerned with the concrete crib wall here. A crib wall is made up of a stretcher, a header, and a filler block. See Figure 10-6.

Place stretchers parallel to the surface of the slope and bury the headers in the earth. The stretchers and headers lock into each other to resist the thrust of the embankment. Use filler blocks at the end of each foundation stretcher. You should grade the earth to a slope of between $1^1/2$:1 and 2:1. Rest the base of a crib wall on at least 5 feet of soil compacted to 95 percent relative density (tested in a laboratory by standard test procedures).

Stretchers are 6 inches deep and 6, 8, or 10 inches wide. Most stretchers are 6 feet long. The 6- and 8-inch-wide units are reinforced with four #3 steel bars. A 10-inch-wide stretcher has six #3 bars. The end of each 6-foot-long stretcher rests on the narrow part of the header. Filler blocks are 6 inches deep and 10 inches wide.

Headers are I-shaped. You lay them horizontally, with their flanges in a vertical position. The flanges keep the stretchers from moving out of place. The web of a header is 5 inches deep and 8 or 10 inches wide. The depth of a flange is 10 inches.

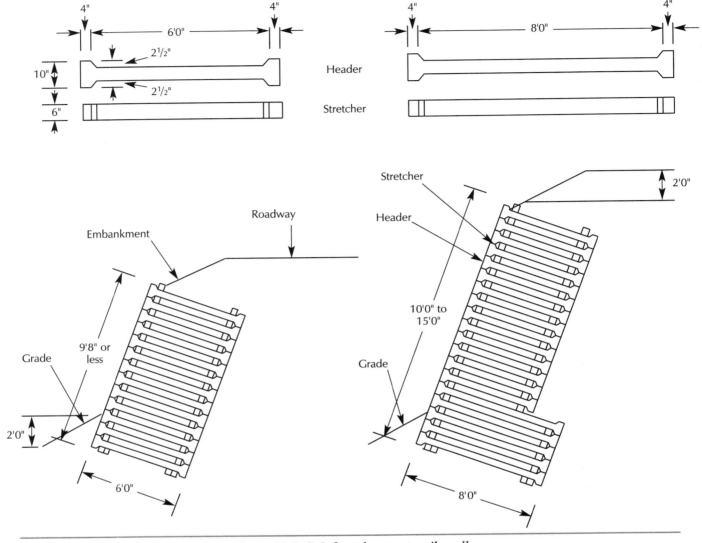

Figure 10-6 *Reinforced concrete crib wall*

A header is reinforced with four #3 longitudinal bars and hooks in the flanges. Headers are 6 and 8 feet long. Use the longer headers in the lower portion of walls between 10 feet and 15 feet high.

Usually, you start a crib wall about 2 feet below grade at the toe of an embankment. Use two pairs of stretchers to support the first header. Set the face of the wall on a batter of 1:6. A batter is the slope of a wall. A ratio of 1:6 means that the wall slopes back 1 foot for every 6 feet of its height.

You can get specifications and details on how to build concrete crib walls at the state highway department, county or city road department, or directly from the manufacturers of precast crib elements.

Water Tanks

Contractors who build industrial or public works may be required to build concrete water storage tanks, waste treatment tanks, or mineral separation tanks. This type of job is basically building high, with curved walls. Most circular tanks with high walls are post-stressed.

A typical five-million-gallon storage tank is 34 feet high, with a diameter of 160 feet. The shell is made of 10-inch-thick concrete cast in curved forms. The wall is reinforced in both faces with #4 bars, with the vertical bars spaced 6 inches on center and the horizontal steel 4 feet on center, staggered. The bars have a 1-inch clearance on the inside surface and a $3/4$-inch clearance on the outside of the shell.

The post-stressing is done after the shell has reached the strength of 3,750 psi. High-strength rods are continuously wrapped and stressed around the shell. The upper 5 feet of the wall is stressed to 35,000 pounds per foot. This is increased uniformly to 216,000 pounds per foot at the base of the wall. The rods are then covered with pneumatic mortar.

The tank wall rests on a concrete ring foundation, 3 feet wide by 1 foot thick. The foundation is cast with a 4-inch-thick concrete floor slab. There's a 4- by $1^1/2$-inch continuous rubber pad between the wall and the foundation. Joints are made watertight with rubber waterstops at the bottom and top of the wall. See Figure 10-7.

Septic Tanks and Clarifiers

Residential areas without a public sewer system are required to have septic tanks. Septic tanks are the simplest form of sewage treatment plant. Such a unit is made of precast concrete and is delivered to the site as a complete tank and lid. In constructing these tanks, high strength concrete is molded in shop-fabricated rigid steel forms that provide close tolerances on all dimensions. A controlled curing process assures attainment of maximum concrete strength. When properly installed and backfilled, the unit will remain in sound structural condition indefinitely.

Most industrial plants are required by the industrial waste control agencies to install clarifiers to separate solids from liquids before the waste enters the public sewer system. A clarifier is very similar to the septic tank.

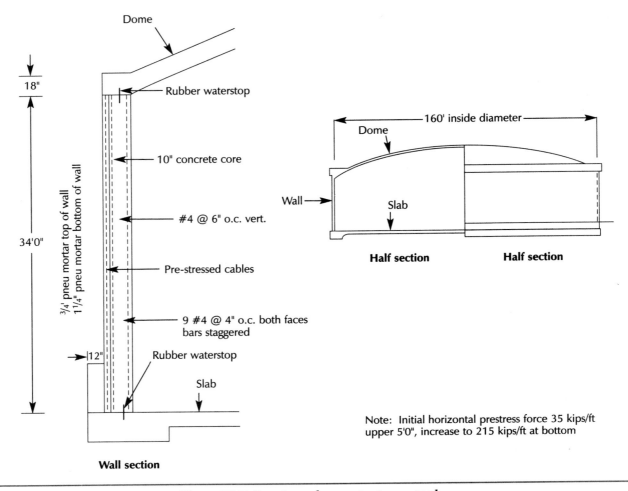

Figure 10-7 *Pre-stressed concrete storage tank*

Septic tanks and clarifiers provide the following functions:

- Separation of floating and settling materials

- Storage and digestion of separated matter

- Biological treatment of suspended particles, changing their character so the discharged water with some solids will filter through the soil without clogging the soil pores

These tanks are available in sizes from 350 to 1,500 gallons. They can be used in batteries to achieve much larger capacities.

The effluent from the septic tank is discharged into the soil either through a leaching pit or through open clay pipes in a leaching field.

Small sewage treatment plants using air pumps to aerate the sewage are also made of precast concrete. These package treatment plants are designed to withstand all normal loads imposed by the earth and liquid pressure. High-strength, high-density concrete is used in these structures, which are reinforced with steel and well cured before installation.

These small sewage treatment systems are available in sizes 1,500 to 15,000 gallons. The larger units have a series of precast concrete chambers butted close together. The blower and motor is above grade in a concrete vault.

The Rest of the Book

Concrete design and construction is a dynamic and ever-changing field, with new materials and methods being introduced every month. Concrete is a composite of many materials, crafts and environmental conditions that goes through a metamorphosis from a stiff slurry to a rigid material. Along the path, it can be affected by workmanship. Stresses are divided between the concrete and reinforcement. All of these uncertainties contribute to the need for safety factors in the design.

The design of concrete structures has developed from rule-of-thumb methods, through graphical methods, to complex mathematical formulas which are continuously being revised and upgraded. That's why using a computer in concrete design is becoming a necessity. Inside the back cover of this book is a CD-ROM with two portions of a powerful concrete engineering program. These are fully-functioning, stand-alone modules, modified to be understood by non-engineers. If you've been able to understand the material in this book, you should have little trouble putting the software to use. You'll find step-by-step instructions in the next chapter. If you are interested in learning about the full program, consisting of 20 modules, turn to page 226 for information about the publisher, Integrated Structural Software, and how to contact them.

The appendices include a checklist you can use on your concrete jobs, and a glossary of terms.

Chapter 11

Using the Concrete Column and Beam Programs

Introduction

Inside the back cover of this book you'll find the *Basic Concrete Engineering for Builders CD*. This disk has two easy-to-use engineering software programs from Integrated Structural Software, Inc. — *Robot RC Calculator* and *Robot Concrete Beam*. These programs will assist you in designing basic concrete elements such as columns and beams. To use these programs, you'll need a computer with at least 48 Mb of RAM running Windows® 95, 98, 2000, NT, or higher and a 17-inch monitor. Up to 18 Mb of free space may be required on your hard drive.

The Robot RC Calculator and Robot Concrete Beam are limited editions of a much larger suite of programs intended for use by registered structural design professionals. But if you've absorbed the information in Chapters 8 and 9, you should have no trouble making good use of Robot RC Calculator and Robot Concrete Beam. For more information about Robot-Millennium analysis and design software, and a quote on other concrete and steel design programs, go to http://www.robot-structures.com/us/index.html. Find a description of Integrated Structural Software, Inc. and its products right after this chapter.

Getting Started

1. Start *Windows*.

2. Insert the *Basic Concrete Engineering* disk in the CD-ROM drive of your computer. After a few seconds the installation process should begin automatically. If it doesn't, click Start, Run, Browse, and navigate to your CD drive, such as D. Double-click on Launch and then on OK.

3. Click Install Products and follow instructions on the screen.

4. This is a two-part installation. Robot Concrete Calculator has to be installed first. When complete, you'll be guided through installation of Robot Concrete Beam.

Install Robot Concrete Calculator

1. Click on Install Robot Concrete Calculator.

2. The RC Calculator ACI Welcome screen will appear.

3. Click Continue to begin unpacking files.

4. Click Next to run the installation setup.

5. Click Next in the information box.

6. Click Next to install the program in the default folder:

    C:\RC Calculator (limited edition).

To choose another folder, click Browse and select the folder and drive you prefer. Then click OK.

7. Click Finish to complete setup of Robot Concrete Calculator.

Robot Concrete Beam

1. Click Install Robot Concrete Beam.

2. Click Next at the Welcome dialog box to begin installation.

3. Click Next to select the default folder:

    C:\Robot Concrete Beam (limited).

    To choose another folder, click Browse and select the folder and drive you prefer. Then click on OK.

4. Click Next to accept the default Program Folder:

    Robot Structural Office 13.5.

    Or, type in the folder name you prefer.

5. Click Finish to restart *Windows*. You must restart *Windows* before you can use these programs.

6. Remove the CD-ROM from your drive.

To open either of the programs, click Start and Programs. You'll find the two newly-installed programs on your program list, probably at the bottom of the list:

    Robot Structural Office 13.5 and RC Calculator ACI (limited version).

The Robot RC Calculator ACI (Limited Edition)

This program follows the American Concrete Institute (ACI) 318-89 Standards, BS 8110, and Eurocodes. It can design and analyze rectangular and T-shaped concrete members under the following load conditions:

- Simple compression (i.e., short columns with concentric load)

- Simple tension (i.e., grade beam ties between piles or caissons)

- Simple bending (i.e., beam with end supports free to rotate)

- Combined axial load and bending (i.e., columns with eccentric load)

- Combined axial load and biaxial bending (i.e., columns with eccentric load in two directions)

- Shear (i.e. short beam with concentrated load near one end)

The purpose of Robot RC Calculator is to find the steel reinforcement required for a defined section and load; then to find the limiting capacity for that section and reinforcement. You can pick an interior or exterior exposure for the member, as this affects the concrete cover over the bars.

Tabs and Buttons of the RC Calculator ACI Program

Notice the four tabs at the top of the dialog box. See Figure 11-1. You have to enter information for your design in each tabbed dialog box to complete an analysis.

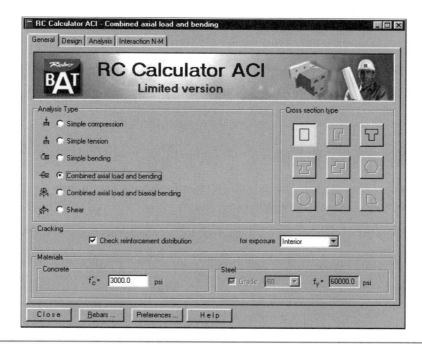

Figure 11-1

At the General tab, choose the analysis type (at the left) and the member cross-section (at the right). Only rectangular and T-shapes are active in this program version. Then select the concrete compressive strength, f'c. Default strength is 3000 psi. To change the compressive strength, select the numbers in the window (such as 3000) and type in the correct concrete strength (such as 3500).

Use the Design tab to calculate the steel area necessary to carry the applied loads for a given section. What you see in this window depends on the type of analysis you selected at the General tab.

Where required, type in the Normal Load (N), the Bending Moment (M), Concrete Cover, Rebar and Stirrup size, and member size (h and b). Default numbers can be changed as just described.

The Analysis tab calculates the maximum loads that may be applied on a selected section and steel reinforcement. Note that the Analysis dialog box is similar to the Design dialog box except that it determines the load that a selected section and reinforcement can carry.

The Interaction N-M tab appears only if you have selected either of the two combined axial load analysis types on the General tab. This dialog box displays complex axial load and bending interaction diagrams on an N-M co-ordinate system. See Help for a detailed explanation of this feature.

Buttons at the bottom of the screen allow you to select the design format:

Rebars — activates a calculator that figures the size, bar number and cross-sectional area for any combination of reinforcing bars. Rebar sizes are given as numbers (#4, #5, etc.).

Preferences — has two tabs, General and Units.

General — The full program lets you select a foreign language, and which code you're working under. In the limited version created for this book, English and the ACI 318-89 code are preset.

Units — lets you select:

- Section dimensions of members in inches (in), feet (ft), millimeters (mm), centimeters (cm), or meters (m).

- Reinforcement cross-sectional area in in^2, mm^2, or cm^2. (Note: these appear in the program as in2, mm2 and cm2.)

- Reinforcement bar diameters and number in feet (ft), inches (in), millimeters (mm) and centimeters (cm).

- Structural dimension (spans or lengths) in inches (in), feet (ft), millimeters (mm), centimeters (cm), or meters (m).

- Forces in pounds (lbs), 1000 lbs or kips (K), Newtons (N), Kip Newtons (KN), or Tons (T).

- Moments are in foot-pounds (ft-lbs), foot-kips (ft-kips), inch-pounds (in-lbs), or inch-kips (in-kips).

- Stresses are in pounds per square inch (psi), or pounds per square foot (psf).

Help — defines the terms and symbols used in RC Calculator ACI, and explains the use of the other buttons on the screen. The Help button may be used as a manual for the Robot RC Calculator ACI program.

Close — exits the program.

Procedure Using the Robot RC Calculator

Go to the **General** tab. The Robot RC Calculator ACI is limited to rectangular and T sections. The basic steps that we follow for sizing any structural member are:

1. Select the Analysis type and Cross-section type.

2. Enter the concrete strength, f'_c, and make any adjustments needed for design considerations. The reinforcing steel strength, f_y, is fixed at 60000 psi.

If you want to check the maximum loading capacity of the section, go into **Analysis**.

Examples for Using the Robot Calculator

Example 1: Design for Simple Compression (i.e. Short Column)

1. Click **General** tab.

2. Click Simple Compression for Analysis type.

3. Check the rectangular Cross-section type.

4. Enter concrete strength, f'_c = 3000 psi.

5. Click **Design** tab.

6. Enter cross-sectional dimensions and concrete cover of member:

   Width, b = 10 in, check Fixed

   Height (Depth, in case of a beam), h = 10 in, check Fixed

   Concrete cover, d_s = 1.5 in.

7. Click **Rebars** button to select the reinforcement:

   Let's try using four #6 bars and see if that works

   Enter 4*#6 =

   The response is 1.77 in² (steel area or As = 1.77 in²).

   Exit the Rebars calculator by clicking on the close button (×) at the upper right of the dialog box.

8. Under Loads (at top), check to make sure the units are in kips. If not, go into Preferences and change them.

   Enter compression force, N = 200 kips.

9. Click **Calculate**.

10. Note "CALCULATIONS COMPLETED!" in the display at lower left. Member can carry a compressive force of 200 kips

    With a theoretical steel area, As = 1.78 in²

    And a reinforcement ratio of p = 1.8%

    Minimum ratio is p_{min} = 1.0%

    Maximum ratio is p_{max} = 8.0%.

11. Now click the **Analysis** tab to verify your design.

12. Click **Calculate**.

The selected column and reinforcement can carry 199.95 kips. This indicates that you can use a 10-inch by 10-inch member with four #6 bars for reinforcement. See Figure 11-2.

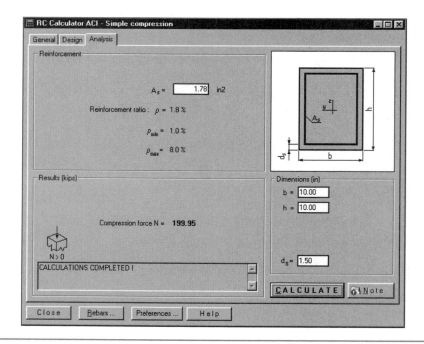

Figure 11-2

Example 2: Beam under Simple Bending

1. Click **General** tab.

2. Click Simple bending for Analysis type.

3. Check rectangular Cross-section type.

4. Check reinforcement distribution.

5. Check Interior exposure.

6. Select Concrete strengths:

Enter concrete strength, f'_c = 3000 psi (default).

7. Click **Design** tab.

8. Enter Factored Moment, M_u = 100 ft-kips.

9. Enter Member type, Beam.

10. Select Rebar size:

   Bottom, or Tension, #6

   Top, or Compression, #6.

11. Select Stirrup size, #4.

12. Enter cross-sectional Dimensions and Concrete Cover of member:

   Width, b = 12.00 in, check Fixed

   Height, h = 18.00 in, check Fixed

   Concrete Cover = 1.5 in.

13. Click **Calculate**.

14. Results appear:

   Required tension (bottom) steel area, A_s = 1.58 in^2

   Required compression (top) steel area, A'_s = 00.0 in^2

   Reinforcement ratio, p = 0.8 %

   Minimum reinforcement ratio, p_{min} = 0.3%

   Maximum reinforcement ratio, p_{max} = 1.6%.

15. Click **Rebars** button to activate calculator:

   Enter 1.58/#6

   The result is 4.

16. Exit Rebars calculator.

17. Click on **Calculate**. Response is once again, Calculation Completed! It's OK to use a 12-inch by 18-inch beam with four #6 bars at the bottom.

18. Click **Analysis** to verify your design.

19. Click on **Calculate**:

> Factored Moment, M_u = 99.19 kip ft.

Your beam design is OK.

The Robot Concrete Beam Program

This program calculates and designs rectangular and T-shape concrete beams. You just enter shape and span (**Geometry**) and the loading condition for the sample beam. It will then calculate the bending moment, shear forces, and the amount of reinforcement you'll need to satisfy ACI requirements. See Figure 11-3.

Headings on RC Beam Program (Limited Edition)

Click on Start, Programs, Robot Structural Office 13.5 and Robot Concrete Beam (Limited Edition). Selections on the menu bar include Materials, Geometry, Loads, Analysis, and Results. Let's look at each of these choices.

Materials — this heading provides a menu for the following items:

▓ Codes (ACI 318-**95** E and U and EC-2)

▓ Materials, such as concrete, longitudinal bars, and stirrups or shear reinforcement

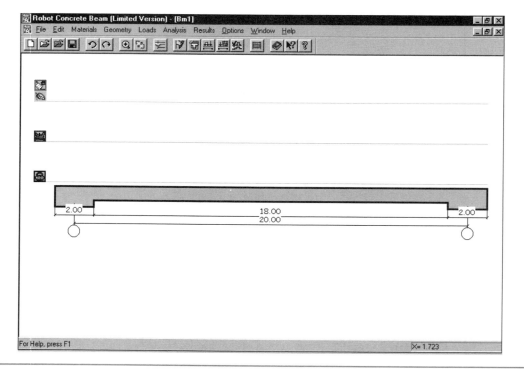

Figure 11-3

- Cover Thickness
- Longitudinal Reinforcement
- Shear Reinforcement
- Design Parameters
- Correction of Internal Forces

Click on another heading or on OK to go back to the first Robot Concrete Beam screen.

Geometry — this heading provides a menu for the following items:

- Topology (beam configuration, add or remove a span)
- Sections (rectangular, T- and L-shape) and Span (center-to-center, clear span)
- Support Conditions
- Spandrel Beam

Loads — this heading selects loading conditions:

- Load Cases (Dead Load, Live Load, Wind 1 and Earthquake 1)
- Beam Self-Weight (weight of rectangular beam or T-beam web)
- Overhanging Plate Self-Weight (weight of T-beam flanges)
- Cursor Load Setting for following conditions:

    Concentrated Load

    Concentrated Moment

    Trapezoidal Load

    Uniformly Distributed Load

    Triangular Load

    Truncated-Triangular Load

    Torsional Moment
- Table Load Setting for Load Cases
- Load Update
- Remove Load
- Remove All Loads
- Remove Loads of Current Type

Analysis — this heading selects the following:

- Complete
- Static

- Load Combinations
- Theoretical Reinforcement Area Calculation
- Reinforcing Bar Arrangement

Results — this heading displays diagrams:

- Moment Diagram
- Shear Force Diagram
- Deflection Diagram
- Theoretical Reinforcement Area Diagram

Options — this heading activates the following:

- Toolbar "Main"
- Toolbar "Geometry"
- Toolbar "Loads"
- Toolbar "Analysis"
- Toolbar "Drawing Flags"
- Status Bar
- Setup for Languages, Directories, Colors and Measurement Units
- Center-to-Center Distance

Example for Using the Robot Concrete Beam Design Program

Calculate the dead and live loads on a two-span beam; then analyze moments, shear forces, deflection and reinforcement diagrams.

When you open the Robot Concrete Beam (Limited Version) program, you'll see the default single-span beam (Bm1).

1. First, add a span to the default beam. Click on **Geometry**, then Topology, and then Add Left Span. See Figure 11-4.

2. Now, select a span by clicking the mouse cursor on the left span section. It will be marked with a red circle to indicate it has been selected. The program numbers the spans from left to right. We've selected Span #1.

3. Click on **Geometry**, then Sections and Span. The Section and Span window appears. You can enter dimensions for Web section depth and width, b and h. Note: Any prior figures remaining may be deleted and new ones inserted. See Figure 11-5.

4. Enter the depth and width of the beam:

    b = 12 in

    h = 20 in

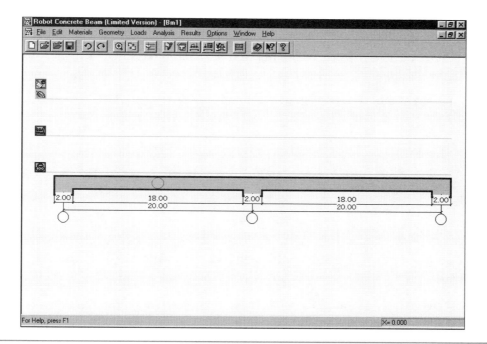

Figure 11-4

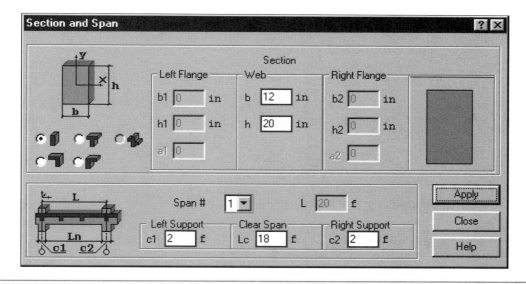

Figure 11-5

5. Select the Span number (#1).

6. Enter the Left and Right Support widths, c1 and c2:

   c1 and c2 = 2f (ft)

7. Enter the Clear Span, or Lc:

   Lc = 18 f (ft)

8. Click Apply, and then Close.

9. Click on the right side of the beam to move the red circle indicator to Span #2.

10. Repeat above steps for Span #2, using

    b = 12 in

    h = 20 in

    c1 and c2 = 2f (ft)

    Lc = 18 f (ft)

11. Click Apply, and then Close.

The geometry for the beam is now defined. Next we'll establish the loads on each span.

1. Click the mouse on the left side of the beam to move the indicator to Span #1.

2. Click **Loads**, then cursor down to Load Cases. A panel appears with Load Cases, including Dead, Live, Wind 1, and Earthquake 1 loads. See Figure 11-6.

3. Click Dead Load. The panel will disappear.

4. Click **Loads** again, and then Table Load Setting. The Loads panel will appear.

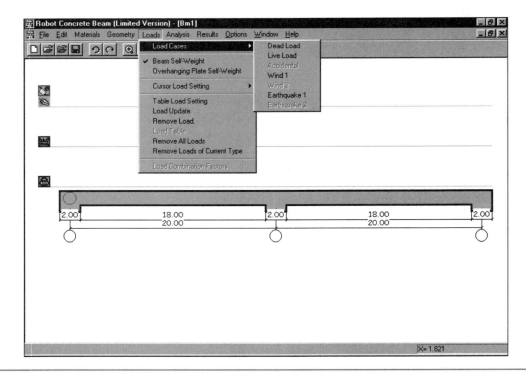

Figure 11-6

5. Select Span #1 and then check the diagram indicating a Uniform load (fourth from top). Note: The dead weight of the member may or may not be automatically calculated when you open up the panel.

6. Enter 5.5 kips/ft for Span #1 in the P1 box.

7. Click Apply and Close.

8. Click on **Loads**, cursor down to Load Cases. Click on Live Loads.

9. Click **Loads**, Table Load Setting. The Load Cases panel will open up. Select Live Load, then click on Choose.

10. The Loads panel will appear again. Select Concentrated Loads (top diagram).

11. Enter a concentrated load of 6.6 kips/ft (P1) at the center of the span, 9.0 ft (a1).

12. Click Apply and Close.

We've established the loads on Span #1. We now need to repeat the process for Span #2.

1. Click on the right side of the beam to move the red circle indicator to Span #2.

2. Click **Loads**, then cursor down to Load Cases. Click on Dead Load. The panel will disappear.

3. Click **Loads** again, and then Table Load Setting. The Loads Cases panel will appear. Select Dead Load, and click on Choose.

4. The Loads panel will appear.

5. Select Span #2 and then check the diagram indicating a Uniform load (fourth from top).

6. Enter 3.0 kips/ft for Span #2 in the P1 box.

7. Click on Apply and Close.

8. Click on **Loads**, then cursor down to Load Cases. Click on Live Loads. The panel will disappear.

9. Click **Loads**, Table Load Setting. The Load Cases panel will open up. Select Live Load, then Click on Choose.

10. The Loads panel will appear. Select Concentrated Load (top diagram).

11. Enter a concentrated load of 4.0 kips/ft (P1). On Span #2 we're going to have concentrated live loads at two points. We'll place the first one at 6 feet. Enter 6.0 ft in the box (a1).

12. Click Apply and Close.

Now we need to repeat the steps above to establish the live load at the second point on Span #2.

1. Click on **Loads**, then cursor down to Load Cases. Click on Live Loads. The panel will disappear.

2. Click **Loads**, Table Load Setting. The Load Cases panel will open up. Select Live Load, then click on Choose.

3. The Loads panel will appear. Select Concentrated Load (top diagram).

4. Enter a concentrated load of 4.0 kips/ft (P1) at 12.0 ft (a1).

5. Click Apply and Close.

   Note the location of the dead and live loads on Spans #1 and #2 in Figure 11-7.

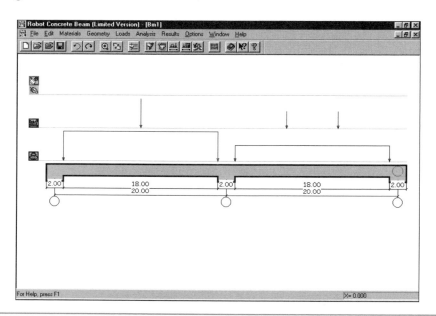

Figure 11-7

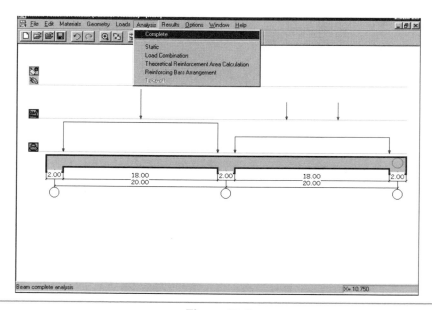

Figure 11-8

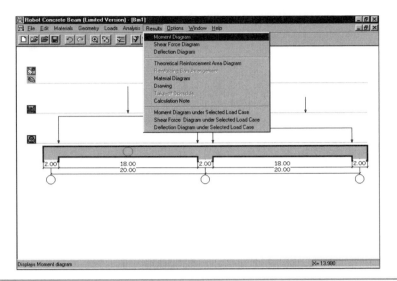

Figure 11-9

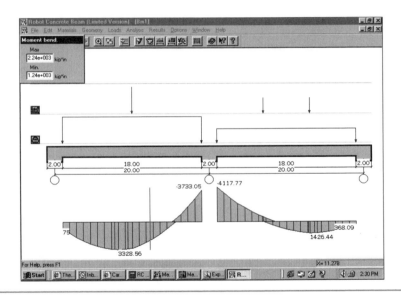

Figure 11-10

6. Click on **Analysis** and Complete to finish. See Figure 11-8.

7. Click **Results**, then on Moment Diagram, Shear Force Diagram, Deflection Diagram, or Theoretical Reinforcement Area Diagram. The respective diagrams will appear on the screen below the beam.

8. Click the mouse cursor anywhere on the diagram and a table will appear at the top of the screen. This table shows the amount of Bending or Shear stress, Deflection, or Reinforcement required (depending on the diagram you've selected). See Figures 11-9 and 11-10.

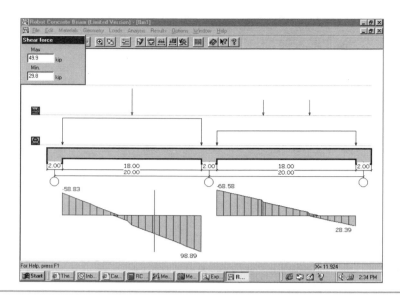

Figure 11-11

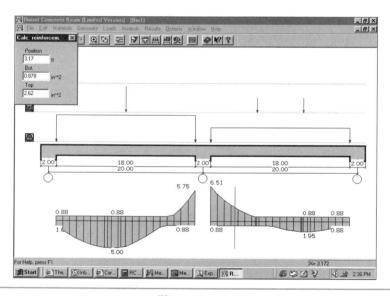

Figure 11-12

9. Holding the left mouse button down, move the cursor across the diagram. The table will indicate the value at any point on the beam for Moment, Shear, or required Reinforcement area. See Figures 11-11 and 11-12.

10. Click File, then Print, to print the results of your calculation.

This completes the analysis of Spans #1 and #2 for Dead and Live Loads for Beam 1. In this example we used the drop-down menus to select items for calculation. You can also use the icons on the various toolbars under the **Options** heading to achieve the same results.

Formulas and Terms Used in Programs

The following ACI formulas were used in the programs.

Basic load combinations are:

1.4D+1.7L

1.2D+1.6L+0.5 S

1.2D+1.6S + 1.8W

1.2D+1.3W+0.5S

1.2D+1.0E

0.9D+/-(1.0E or 1.3W)

0.75 (1.4D+1.7L+1.7W)

Where:

D = Dead load

L = Live load

W = Wind load

E = Earthquake load

S = Snow load

H = Soil load

F = Liquid load

Terms Used in These Programs

Charge	=	load
Accidental load	=	short term loads
Topology	=	number and lengths of beam spans
Geometry	=	cross-sectional dimensions of member
Self-weight	=	dead load of member
Design	=	calculation of theoretical steel area as a function of the applied loads
Analysis	=	calculation of the load capacity of a given cross-section and reinforcement
Bending flexure	=	deflection

Conversion Factors: English and Metric

Multiply	by	to obtain
Length		
inch	25.400	m
foot	0.304800	m
yard	0.914400	m
mile	1.609347	km
mm	.039370079	inch
cm	0.01	m
m	100.00	cm
km	0.621370	mile
m	39.37	inch
Area		
in²	645.160	mm²
ft²	0.92903	m²
Mass		
kg	2.204622	lb
kg	.001102311	short ton
Force		
pound-force	4.44822	newton (N)
newton	0.224809	lb-force (lbf)
kip	1000.00	pounds (lb)
Bending moment		
pound-force-inch	0.112985	newton-meter (N-m)
pound-force-foot	1.355818	newton-meter (N-m)
newton-meter	8.850748	lb-force-inch (lbf-in)
newton-meter	0.737562	lb-force-foot (lbf-ft)
Pressure stress		
lb-force per sq inch	6.894757	kilopascal (kPa)
kilopascal	0.145038	lb-force per square inch (lbf/in²)

Summary

The RC Calculator allows the calculation and pre-engineering of reinforced concrete sections according to ACI 318-89 Code. All loads are understood as "factored loads," that is, they are the resultant combinations of factored loads, applied at the center of gravity of the cross-section. Section resistance, calculated in the Analysis case, is a "factored resistance" or the maximum factored load which may be applied to the section. The program calculates "theoretical" steel area, which may be rounded-up when a specific bar size is chosen.

Information on Robot Programs can be obtained from:

Integrated Structural Software, Inc.
155 Dorchester Way
San Francisco, CA 94127
Toll free: 888-477-8491
Fax: 415-682-8490
Email: mo@issstanford.com
Web site: www.robot-structures.com

These programs are mainly based on the American Concrete Institute (ACI) *Manual of Concrete Practice 1995, Part 3, Use of Concrete in Buildings — Design Specifications, and Related Topics,* and specifically to ASTM Standards ACI 318-95. Its latest edition can be purchased at your local construction bookstore, or write:

American Concrete Institute
Box 19150, Redford Station
Detroit, MI 48219

You should use ACI in conjunction with your local building code, the *Uniform Building Code* (UBC), the *BOCA National Standard Building Code,* the *Standard Building Code* (SBC), or the *International Building Code* (IBC).

Disclaimer

Robot RC Calculator and **Robot Concrete Beam** are designed to handle concrete engineering calculations. The user has the entire responsibility to make design choices; the programs will show you the consequences. You'll quickly get a feel for the general range of what's practical and what isn't.

Please note that no computer program can take the place of professional advice. Although these programs have been prepared with reasonable care and use standard engineering formulae, it may be very difficult to determine exactly what loads the column or beam will actually be carrying, especially for complicated rooflines, or unusual floor framing.

There are many factors involved in the actual design of concrete columns and beams, especially longer members, connections, cantilevers, and concentrated loads. A computer program can't take the place of professional experience and judgment. If you have any questions about the results of the program, please consult your local building inspector, architect, or engineer. Neither Craftsman nor ISS intend to give professional advice and they accept no responsibility or liability in any manner for how the programs may be actually used.

Integrated Structural Software, Inc.
San Francisco, CA
415-682-2205 and Toll free: 888-477-8491
Fax 415-682-8490
info@issstanford.com
www.robot-structures.com

Integrated Structural Software, Inc. produces and markets ROBOT Millennium, a general-purpose finite element analysis and design program for the structural engineering community. The original version of this program was written in 1982 by Andre Niznik in France. ISS set up an office in San Francisco and first offered ROBOT to the U.S. in the early '90s.

In this book, ISS has included a working version of their RC beam design and RC calculator modules.

I.S.S. Inc. provides a top-quality solution to assist engineers in the modeling, analysis and design process of multi-material structures.

ROBOT Millennium for Windows 95/98/NT/2000 or higher is suitable for both simple and complex structures from 2D frames to 3D shell models. It covers a broad range of analysis including advanced options and design capabilities.

Main Features are:

- ✓ 2D/3D frame and truss analysis

- ✓ Plates and shells finite elements analysis with automatic and powerful meshing

- ✓ Various analysis types: linear, non-linear, static, dynamic, modal, spectral...

- ✓ Advanced options such as cables, P-delta, buckling, seismic...

- ✓ Bridge analysis with moving loads

- ✓ Design code checks for steel (per ASD-LRFD) and concrete (per ACI) structures

- ✓ International units, section databases and design codes

- ✓ Material take-off and cost estimation

ROBOT strengths are both the power and the flexibility of its graphical user interface. Thanks to the smart options outlined below, engineers can increase their productivity:

- ✓ Powerful modeling process using libraries of structures

- ✓ Import/Export to CAD systems: DWG, DXF, CIM Steel formats....

- ✓ Complete and immediate post-processing using tabulated or graphical output display

- ✓ Customization of calculation reports and generation of documents in any Windows application

- ✓ Direct link of ROBOT input and output with other applications through Microsoft COM technology

ROBOT is a 100% Windows application and the only structural software including COM Technology. It allows seamless interface with other applications like in-house programs and CAD software. For more information, please visit www.robot-structures.com or call ISS.

Concrete Checklist

Forms

Here are some general rules for designing wood forms:

- ❑ Use stock sizes and lengths of lumber.
- ❑ Use as few lengths of lumber as possible.
- ❑ Use as few units as possible, but don't make the units too heavy.
- ❑ Design the forms for easy stripping.
- ❑ Try to use such units as wall panels, floor panels, and beam and column forms that you can reuse.
- ❑ Make bevel cuts and keys so you can release forms with little prying.

Here's a checklist to use when you design concrete wall forms:

- ❑ Find out what materials are available.
- ❑ Determine the rate of delivery of concrete to the job site (in cubic yards).
- ❑ Calculate the area to be enclosed by the concrete (in square feet).
- ❑ Calculate the rate of pour in the forms (in vertical feet per hour).
- ❑ Estimate the air temperature at the time of placement.
- ❑ Determine the maximum concrete pressure in the forms (in pounds per square foot). See Figure 5-9 on page 90.
- ❑ Figure the maximum stud spacing (in feet). See Figure 5-2 on page 81.
- ❑ Calculate the unit load on the stud (in pounds per linear foot).
- ❑ Find the maximum wale spacing from the strength of the studs and the lateral pressure (in feet).
- ❑ Figure the uniform load on the wales (in pounds per linear foot).
- ❑ Determine the tie wire spacing based on the wale size and on wire strength.

- ❏ Compare the maximum tie spacing with the maximum stud spacing.

- ❏ Determine the number of studs and wales for one side of the form.

- ❏ Brace wall forms, in either direction, against wind. Check with your local building department for wind loads in your area.

- ❏ Attach the sheathing to the studs with as few nails as possible to make it easier to disassemble the forms. You may need extra nails on gang forms. Use 6d nails for $^{23}/_{32}$-and $^3/_4$-inch plywood.

- ❏ Verify forms are true, tight, oiled or wetted and clean of debris.

- ❏ Make sure forms remain in place for proper length of time.

- ❏ Check form cleaning and repairing.

Reinforcement

- ❏ Make sure that reinforcing steel is free from loose scale and is properly tied and supported.

- ❏ Check the bending and cutting list before cutting, bending, bundling and tagging.

- ❏ Confirm you're using the right wire mesh and welded fabric, and plain and deformed bars

- ❏ Check the placing and tying in place, bar spacers and supports and tie wire, loops and anchors.

Placing Concrete

Review this checklist before you pour any concrete:

- ❏ Make sure any excavation you did for the foundation is dry and free of water, snow, or ice.

- ❏ Don't place concrete on porous ground.

- ❏ Sprinkle or seal semiporous subgrade to prevent suction that could make a mix too dry and stiff. Just make dry soil slightly moist so that it won't draw water from the fresh concrete. Otherwise, the concrete may get too stiff to work or place properly.

- ❏ Make sure you've put up stable and complete formwork.

- ❏ Make sure you've put the reinforcing steel and other imbedded items, such as expansion joints and anchor bolts, securely in place.

❏ Set the openings and sleeves for ducts, pipes, and conduits and have them checked by the job superintendent.

❏ Have the forms and reinforcing bars inspected and approved by the building inspector.

❏ Remove and clean all dirt from contact surfaces so that the fresh concrete will bond well. Clean the top of footings before you pour the pedestals, and brush off all construction joints in hardened concrete slabs and walls before you extend the slab or wall with fresh concrete.

❏ Make sure all equipment is clean. You should clean it at the end of each operation or day's work.

❏ Remove any hardened concrete and foreign materials from the conveying equipment.

❏ When freezing and thawing might occur, check limiting temperature for protection from drying or too rapid cooling.

❏ During hot weather, check for prewetting of aggregates and contact surfaces.

❏ When ready-mix concrete is delivered, check:

Elapsed time

Number of revolutions of drum

Mix

Separation during delivery

Unauthorized adding of water

Check shipping tickets for proper mix and water-cement ratio

Verify proportions for code compliance

Use this checklist as you pour concrete:

❏ Mix the concrete thoroughly with a rod, or preferably a mechanical vibrator.

❏ Don't dump concrete directly from a mixer into a bucket. This throws the heavier and larger rock to one side.

❏ Place concrete continuously in layers.

❏ If you use a buggy to pour concrete into a formed wall, unload it in a concentric fashion using a vertical chute.

❏ Pour the concrete in even horizontal layers of uniform thickness, 6 to 24 inches in depth. Make each layer 6 to 12 inches thick for reinforced members, and up to 18 inches thick for mass concrete work.

❏ Be sure to place each layer before the layer below it sets up.

❏ Pour concrete as close as possible to its final position.

❑ Don't place concrete in large quantities at one point and let it run or be worked a long distance in the forms. This may segregate the material.

❑ Pour concrete at the far end of slabs, against the concrete you just placed.

❑ Avoid bouncing a mixture off one side of a form. This makes the mixture separate.

❑ If you can't get a mixer truck up to your forms, pump concrete from the truck to the forms.

❑ Dump concrete from a transit-mix truck by chute into a hopper. Then pump the mix through a hose mounted on a specially-designed truck with an articulated boom.

❑ Don't let concrete drop freely more than 3 or 4 feet. In thin sections, use rubber or metal drop chutes. In tall, narrow forms, you can place the concrete through openings, or windows, in the sides of the forms. Don't let a chute discharge directly through an opening as this may cause the mix to separate. Use a rectangular chute with a steel hopper at the top for placing concrete in narrow forms.

❑ Keep fresh concrete from bleeding by using low-slump, air-entrained concrete that has adequate cement and properly graded sand. Also, don't float or trowel concrete until it's hardened enough so the water and fine material in it don't come up to the surface.

❑ Don't spread dry cement on a wet surface to take up excess water. That can form a paste that will break off, or cause dusting and scaling. It's better to wait for all water to evaporate or be removed. You can drag a rubber garden hose over the surface to wipe the water off. Another way is to use fans or blower type heaters to dry off the water.

Foundations

❑ Is there a soils report? If not, use the soil classification and bearing values in the building codes when you make a preliminary design.

❑ Check the type of soil classification used for nearby building permits.

❑ Is there a contour map of the original grade? Compare the original contours with the final contours. This will tell you whether the site has been filled or cut.

❑ If there is a fill, what's it like and how far does it go? If the fill is over 3 feet deep, check with the grading department on whether the fill was certified. If it was, ask them what the allowable soil pressure is. If it wasn't, don't assume that it's good enough to support the building. Extend the foundations to natural grade. For uncertified fill deeper than 3 feet, have a soils engineer make an investigation and recommend what type of foundation you should use.

❏ Find out if there are underground or aboveground obstructions. Adjacent buildings can get in the way of any excavating you'll have to do.

❏ Find out from the building department how deep the frost line is.

❏ How deep is the water table? A high water level can flood footing excavations and caisson holes. Find out if nearby builders found ground water. If there's a possible high water table, hire a soils engineer to bore some test holes.

❏ What are the elevations of finish grade, column foundations, and building floor? You'll need at least a 6- to 8-inch high foundation above finish grade.

❏ What is the soil bearing value at the depth of the footing? If you didn't get a soils engineer's report of the soil, dig a trench or pit to find out if there's any fill.

❏ If there's a slope on part of the property, or adjacent to the property, check the angle of slope. A cut or fill slope shouldn't be steeper than 2:1 (horizontal to vertical) unless the slope was approved by the building department or a soils engineer.

❏ Check with the grading department on how close you can build to the edge of a fill or cut slope.

❏ Check the building code to find out how far below the natural or finish grade the footing has to be. In expansive soil, such as adobe, you have to put in a deeper footing to keep moisture from reaching the bottom of the footing.

❏ If there are cut or fill slopes, check with the neighbors and grading department if there have been landslides, soil-slumps or mudflows in the nearby lots.

❏ Inspect for visible cracks in adjacent buildings and surrounding ground. These signs are an indication of expansive soil, deep-fill settlement, or soil-bedrock downslope creep.

❏ Check for evidence of burrowing rodents. This may be an indication of loose fill. The combination of loose fill and a network of rodent holes makes the slope vulnerable to saturation and slump-type failure.

❏ In landslide, questionable or hazardous areas, no building or grading permit is normally issued until the grading department gets reports from a soils engineer which certify the safety of the area.

❏ Don't allow drainage over a slope unless it's controlled by drainage devices such as pipes or channels. Drain rainwater toward the street and away from the slope if possible.

❏ Plant cut and fill slopes to prevent erosion. Check with the grading department for the recommended types of plants for slope erosion control. Install a sprinkler system.

❑ If a building site is on a hillside, check the location of natural water courses. Install debris basins, debris fences, or other devices in the water courses to protect the site from mud slides.

❑ Don't do any grading on hillside lots during the rainy season.

Slabs

Verify these for slabs-on-grade:

❑ Joints are as shown on plans, including:

Construction joints

Expansion joints

Control joints

❑ Steel reinforcement placement, including:

Size of bars

Location of bars

Spacing of bars

Splices

❑ Temperature reinforcement for one-way slabs.

❑ Preparation for curing

Consider these factors when you select any slab that you don't cast on grade:

❑ How will the slab be used?

❑ What are the loads on the slab — light or heavy, uniform or concentrated?

❑ Is the span of the slab long or short?

❑ How will the slab be supported?

❑ Are there any obstructions to the columns, beams, and girders?

For a cantilevered concrete balcony or exposed elevated concrete slab:

❑ Is the slab too thin to let you place the steel reinforcing bars properly?

❑ Is there enough concrete cover over the reinforcing bars?

❑ Are there aluminum handrail posts close to steel reinforcement bars? This can cause electrolysis that will make the bars corrode.

❑ Is the concrete so porous that moisture will get in and corrode the reinforcement bars?

❑ Are the construction joints sealed properly so moisture can't reach the reinforcing?

❑ Is there too much chloride in the concrete so the reinforcing bars will rust?

❑ Is there enough waterproof coating on the concrete so the bar won't corrode?

❑ Is there enough slope for drainage so water won't pond up and penetrate the concrete cover?

Beams

You should keep in mind that you're designing against potential failure. When designing a beam, ask yourself the following questions:

❑ How many stirrups and diagonal bars do I need near the end of the beam to keep it from cracking?

❑ Do I have enough reinforcement in the beam to keep it from cracking or being crushed by a heavy load? Remember, you need reinforcement at the bottom of a simple beam, and near the top at the supports of a continuous beam.

❑ Do I have enough lateral support for the beam to keep it from buckling?

❑ Is the beam wide enough to keep it from buckling under a heavy load?

❑ Are the bars and the concrete full bonded together?

❑ Have I lapped the bars properly so they act like a continuous bar? Figure 9-6 on page 190 shows the right and wrong ways to lap bars. See Figure 9-7 on page 190 for minimum lap for bars.

Walls

❑ Check reinforced concrete walls for:

Ratio of height to thickness

Maximum height

Minimum thickness

Minimum reinforcement

Special reinforcement shown on plans

❏ Check tilt-up precast walls for:

Location of pickup points

Strength of concrete

Separation of precast sections from casting bed

Adequate lifting equipment and accessories

Competent operators

Satisfactory base anchorage

Safety provision for bracing

Adequate connections

❏ Verify that concrete for all walls is properly compacted and vibrated.

Testing

❏ Take specimens of ready-mix concrete for slump and strength tests.

❏ Confirm that concrete materials are identified as complying with appropriate standards.

❏ Make sure any cuts made in the finished concrete do not impair structural strength.

❏ Keep necessary records of batching and mixing, placing and curing.

❏ Check if adequate borings or excavations are made to verify foundation design.

This form is available for free download at:

www.craftsman-book.com/cbcstore/prodpages/info/bcb/checklist.htm

Glossary

A

Abram's Law: A rule stating that the ratio of water to cement determines concrete strength.

Acceleration (earthquake): Rate of ground movement as compared with the acceleration of a falling object.

Accelerator: An admixture used to increase early strength gain in concrete.

Accepted engineering practice: Design and construction that conforms to accepted principles, tests, or standards of nationally recognized technical and scientific authorities.

Acrylic (wood): A wood plastic composite, produced by impregnating wood with an acrylic monomer polymerized by ionizing radiation or other techniques, that's used in some formwork.

A.D.: Air-dried lumber.

Addenda: Documents issued before the bid opening which clarify, correct, or change bidding documents or contract documents.

Adhesive: A substance that holds materials together by surface attachment.

Admixture: Material other than water, aggregate, or cement, used as an ingredient of concrete.

Aftershock (earthquake): An earthquake of less intensity than an initial earthquake.

Aggregate: Granular material such as sand, gravel, crushed rock and iron blast furnace slag, which, when used with portland cement, forms concrete.

Aggregate, coarse: A concrete component with a particle size over $1/4$ inch.

Aggregate, fine: A concrete component with a particle size less than $1/4$ inch.

Aggregate, lightweight: An aggregate with a dry loose weight of less than 70 pounds per cubic foot.

Aggregate/cement ratio: Weight of aggregate divided by weight of cement.

Agreement: Document signed by an owner and contractor covering the work to be performed.

Air-dried (A.D.) lumber: Lumber used in formwork that has been stored in an open area to dry naturally. Also see *Seasoning*.

Air-dry weight: Unit weight of a lightweight concrete specimen cured for 7 days with no moisture loss or gain.

Allowable span: Greatest horizontal distance permitted between supports.

Allowable stress: Amount of force per unit area permitted in a structural member.

Allowable stress increase: Percentage increase in the allowable stress based on the length of time that the load acts on the member.

Anchor: Device used to secure formwork to previously placed concrete of adequate strength, normally embedded in the concrete during placement.

Anchor bolts: Steel bolts embedded in concrete that hold a building or structure to the foundation.

APA: American Plywood Association. They represent most of the plywood manufacturers for the purpose of research, quality, and promotion.

Architect: Person licensed by the state and charged with the design and specifications of a building.

ASTM: American Society of Testing and Materials.

Axial force: A push (compression) or pull (tension) acting along the length of a member, usually measured in pounds.

Axial stress: Load or force divided by the cross-sectional area of a member, usually expressed in pounds per square inch (psi).

B

Back (plywood): Back veneer of a plywood panel that is normally of lower quality than the front veneer.

Backshore: A shore placed under a concrete slab or beam after the formwork and original shores have been removed.

Bag (sack): A quantity of portland cement that weighs 94 pounds.

Bar: Metal rod used to reinforce concrete. Also called *reinforcing steel, reinforcing bars, or rebar.*

Bar chair: Support for reinforcement bars during concrete placement.

Barrel (cement): Weight measure of portland cement (four bags or 376 pounds).

Base shear (earthquake): Total horizontal seismic force exerted at the top of a foundation.

Basic wind speed: Probable fastest wind speed measured 33 feet above the ground in a flat open area.

Batch: Quantity of concrete mixed at one time.

Batter: Inclination from the vertical.

Batterboards: Elevated horizontal boards set at the corner of a building used to establish the locations of the corners and elevation of the building foundation wall.

Beam: A horizontal load-bearing structural member.

Beam bottom: Soffit or bottom form for a concrete beam.

Bearing wall: A wall that supports a load in addition to its own weight.

Bending moment: Measure of bending effect due to a load acting on a member, which is usually measured in foot-pounds, inch-pounds, or inch-kips.

Bending stress: The force per square inch of an area acting at a point along the length of the member, resulting from the bending moment applied at that point, usually expressed in pounds per square inch (psi).

Bid: Written proposal submitted by a bidder stating the prices for the work to be performed.

Bleeding: Water on cement surface due to settlement of solids in the mix.

Blind nailing: Driving nails so that nail heads are concealed.

Blocked diaphragm: A diaphragm in which all sheathing edges are supported by framing members or blocking.

Blocking: Small wood pieces installed between studs, joists, or other members to prevent buckling of formwork.

Board foot: Unit of measure equivalent to a board 1 foot square and 1 inch thick.

Boards: Lumber that is 2 or more inches wide and 1^1/$_2$ inches (or less) thick.

Bonded tendon: Prestressing tendon that is bonded to concrete either directly or by grouting.

Borer holes: Wood voids made by grubs, worms, and wood-boring insects.

Bottom plate: See *Sill plate.*

Boundary element (diaphragm): The edge condition of a horizontal or vertical diaphragm.

Bow: Lumber distortion parallel to the grain.

Box-out: Opening or pocket formed in concrete.

Brace: Load-bearing member installed diagonally.

Bridging (or cross-bridging): Diagonal bracing placed between joists and other members supporting forms for elevated slabs to keep them from buckling.

Buck: Framing around an opening in a wall.

Buggy: Two-wheel or motor-driven cart used to carry small quantities of concrete to forms.

Bug hole: Void in the surface of formed concrete caused by adhering air or water bubble not displaced during consolidation.

Builder's level: Surveying instrument to control horizontal planes.

Building paper: Heavy paper used to waterproof walls and roofs.

Building permit: Document issued by the building department certifying that plans have been approved for construction.

Built-up member: Single structural wood member made from several pieces fastened together.

Bulk cement: Cement delivered in large quantity.

Bulkhead: Partition built into wall forms to end each concrete pour.

Butt joint: Straight wood joint with an interface perpendicular to the grain.

C

Camber: Predetermined curve set in a beam or slab to make up for the sag that will occur when the member is loaded.

Cant strip: See *Chamfer*.

Catwalk: Narrow elevated walkway.

Cellular concrete: Lightweight concrete made of portland cement, lime/silica, or lime/pozzolan.

Cement mason: A craftsman who smoothes and finishes surfaces of poured concrete walls, floors, and sidewalks.

Centering: Temporary supports placed under arches, shells, and space structures that are removed or lowered as a unit to prevent destructive stresses on a structure caused by unequal support.

Chalk line (snap line): Spool-wound string encased in a chalk-filled container that is pulled taut across a surface, lifted and snapped directly downward so that it leaves a straight chalk mark.

Chamfer: Beveled edge formed in concrete by a triangular strip of wood (chamfer strip) placed in a form corner.

Change order: Written order to the contractor, signed by the owner, authorizing him to add, delete, or revise work.

Check (lumber): Lengthwise separation of wood used in formwork that usually extends across or through annual growth rings.

Cleanout: Opening at the bottom of forms that allows access for removing refuse.

Clear span: Horizontal distance between the interior edges of the supports for a beam or truss.

Climbing form: A form which is raised vertically for succeeding lifts of concrete in a given structure, usually supported on the anchor bolts or rods embedded in the top of the previous lift. The form is moved only after the entire lift is placed and (partially) hardened. This should not be confused with a *slip form* which moves during placement of the concrete.

CLR: Abbreviation for *clear* when used to describe minimum distance between reinforcing bars or between the bars and concrete surface.

Column (post): (1) A vertical load-bearing structural member. (2) A member with a ratio of height to least lateral dimension of 3 or greater. Used primarily to support axial loads.

Combined stress: Combination of axial and bending stresses acting on a member simultaneously.

Common nail: Steel wire nail.

Compression: Force that tends to crush a structural member.

Concentrated load: Load centered at a given point.

Concrete: Mixture of portland cement, fine aggregate, coarse aggregate, and water, with or without admixtures.

Concrete, lightweight: Low density concrete made by using lightweight aggregate. Usually weighs about 115 pounds per cubic foot.

Concrete mix designer: Technician who determines proper quantities of aggregates to meet specifications of a concrete mix.

Concrete mixer operator (or batch plant operator): Technician who tends mixing machines to mix sand, gravel, cement, and water to make concrete. Also cleans and maintains mixer.

Concrete, normal weight: Concrete made with natural crushed rock and sand, usually having a unit weight of 135 to 165 pounds per cubic foot.

Construction documents: All the written, graphic, and pictorial documents describing the design, location, and physical characteristics of a building that are necessary for obtaining a building permit.

Construction joint: The surface where two successive placements of concrete meet; frequently there's a keyway or reinforcement across the joint.

Control chemist (batch plant): Technician who makes periodic chemical analyses of concrete mix to make sure it is uniform in content and meets specifications. He also keeps records and makes reports.

Corner brace: Member installed diagonally at the corners of wall forms.

Cross-bridging: See *Bridging*.

Cross section (beam): Section taken through a member perpendicular to its length.

Crown: Upward bow in a horizontal structural member.

Crowning: Installing a horizontal member with its crowned edge up.

CRSI: Concrete Reinforcing Steel Institute.

D

d: Abbreviation for "penny" used to designate nail size.

Dead load: Permanent weight of a building structure, including equipment.

Decay: Deterioration of wood due to fungi.

Decenter: To lower or remove centering or shoring.

Decking: Sheathing material used for deck or slab soffit forms.

Deflection (beam or truss): Amount of sag in a horizontal structural member. Usually expressed as a ratio of the amount of deflection to the span of the beam.

Deformation: Changes in a member's shape, such as shortening, lengthening, twisting, buckling, or expanding. Also called *strain.*

Delamination (plywood): Separation between plies, normally due to moisture.

Dense select structural: High-quality lumber used in formwork, relatively free of characteristics that impair its strength or stiffness.

Depth: Board dimension measured parallel to the direction of the principal load on the member.

Deputy inspector: Specially approved building inspector. Also called a *special inspector.*

Design earthquake: 90 percent probability of design not being exceeded in 50 years.

Design load: Total load that a structural member is designed to support.

Diaphragm: Horizontal or nearly horizontal system designed to transmit lateral forces to the vertical structural members of a building.

Diaphragm chord: Outer edges of a horizontal or vertical diaphragm.

Diaphragm strut: Compression or tension member that transfers horizontal seismic loads to a diaphragm.

Diaphragm (vertical): See *Shear wall.*

Dimension lumber (standard dressed lumber): Lumber that is 2 to 5 inches thick and up to 12 inches wide, including joists, rafters, studs, planks, posts, and small timbers.

Drift: Horizontal movement of a building caused by an earthquake.

Dry mix: All the ingredients of a concrete mix without moisture.

Dumpy level: Surveyor's instrument used to control horizontal planes.

Dust-on: Casting dry cement on a concrete surface to absorb bleed water.

E

Earthquake magnitude: A system that measures the effect or intensity of an earthquake at a specified point in terms of a series of levels. This may be from I to XII in the Mercalli Scale, measured by the observed severity of damage, or from 1 to 8 in the Richter Magnitude Scale, based on the intensity of movement at the earthquake's epicenter by various seismographs.

Edge nailing (plywood sheathing): Series of nails along edge of each plywood panel.

Elastic limit: Amount of stress that concrete can't recover from.

Elasticity: See *Modulus of elasticity.*

Elephant trunk: Sectional metal tube for placing concrete in high forms.

End nailing: Nails driven into the end of a board.

Engineered wood: Specially-designed structural member or assembly that is usually built off-site.

Entrapped air: Air present in concrete that is not added purposely, as by adding an air entraining admixture.

Equilibrium moisture content: Moisture content of wood that is in balance with the relative humidity.

Exposure (wind): Description of the terrain surrounding a building and the highest wind velocity with regard to wind exposure.

F

Fabricated structural timber: Engineered wood member, including sawn lumber, glu-laminated timber and mechanically-laminated lumber used in formwork.

Face (plywood): Side of a plywood panel that is of higher veneer quality when front and back are of different veneer grades.

Factor of safety: Allowable unit stress based on judgment of a competent authority, the risk involved, consistent material quality, and loading condition control.

False set: Premature stiffening of concrete. Concrete can be made plastic again by vibrating.

Falsework: Temporary structure erected to support work in progress, such as shores or vertical posts supporting formwork.

Fastest mile wind speed: Highest sustained average wind speed on a mile-long sample of air passing a fixed point.

Fat mix: Cement mixed with a high cement factor.

Fault (earthquake): Zone of weakness in the earth's crust allowing movement between adjacent crust blocks.

FBM: Feet board measure.

Fiber-reinforced gypsum panel: Construction material made of gypsum slurry and plant fibers formed into large sheets.

Fiberboard: Construction material made of wood or other plant fiber compressed into large sheets.

Final set: Concrete that has hardened enough to resist penetration of a weighted test needle.

Fire-rated: Construction material that is tested and shown to be fire-resistant for a given period of time.

Fire-resistive construction: Construction where the structural frame is protected by covering it with plaster or gypsum wallboard.

Fire-resistive rating (fire resistance): Time (in hours or fractions) that a material will withstand exposure to fire. Based on ASTM testing procedures.

Fire-retardant wood: Any wood product pressure-treated with chemicals that shows a flame-spread index of 25 or less when tested according to ASTM E-84.

Flame spread index: Rate at which a flame spreads over a surface.

Flange (built-up beam): Horizontal members of an I-beam or box beam.

Flexural strength: Resistance to bending stress.

Floor girder: Beam that supports floor joists.

Flying forms: Large mechanically handled sections of formwork; frequently includes supporting truss, beam, or scaffolding frames completely unitized.

Fog curing: Storage of concrete test samples in a moist room, under a controlled temperature, maintained damp by a fine fog-like spray.

Force: Push or pull exerted by one object on another.

Force diagram: Graphic representation of forces as they interact within a structural system.

Force, earthquake: Acceleration from rest to a velocity of a building resulting from sudden earth movement.

Formwork: Total system built to contain freshly-placed concrete, including sheathing, supporting members, hardware, and bracing.

Foundation: Brick or concrete support wall that a house or building sits on.

Full-size lumber (sawn lumber): Undressed or rough lumber.

G

Gang nails: Light-gauge metal plates used as connectors for wood members.

Ganged forms: Prefabricated form panels joined to make a larger unit for efficiency in erecting, stripping, and reuse.

Gin pole: Nearly vertical wood post used to hoist heavy material and equipment.

Girder: Major horizontal structural member that supports secondary beams, joists, or rafters.

Girt: Horizontal member used to support wall siding.

Glu-laminated (glu-lam) timbers: Structural members made of wood, plywood, or both, bonded together with adhesive.

Grade-marked lumber: Lumber that has been inspected and stamped showing the specie and quality of the wood.

Green lumber: Freshly sawn, unseasoned, or undried wood.

Gusset: Small piece of wood, plywood, or metal attached to the corners or intersections of a frame to add strength and stiffness.

H

Hold-down connectors: Steel anchor straps used to bolt a shear wall to a foundation.

Honeycomb: Void left at the formed concrete surface revealing coarse aggregate.

Horizontal seismic force: Reaction of a building or structure to the movement of the ground during an earthquake.

I

Identification index (plywood): See *Span rating*.

Intermediate nailing (sheathing): Series of nails within the interior of plywood panels.

J

Jack: Mechanical device used to adjust the elevation of forms or form supports.

Jack shore: Telescoping adjustable single-post metal shore.

Jack truss: a truss used to support another truss and eliminate a post.

Joists: Horizontal structural members that support a floor or ceiling.

Journeyman: Tradesman with the experience required to complete any task without supervision.

K

K.D.: Knocked down.

Keel: Oil crayon used to mark the locations of framing members.

Kerf: Notch or cut in a beam.

Kiln: A long rotating drum in which crushed limestone is baked, then ground into a powder forming portland cement.

Kiln-dried lumber: Wood seasoned in a special chamber using artificial heat.

Kip: Unit of force representing a thousand pounds.

L

L-head: Top of a shore that has a braced horizontal member projecting on one side, forming an inverted L-shaped assembly.

L-shore: Shore with an L-head.

Lagging: Heavy sheathing used for underground work to temporarily support earthen walls.

Lateral brace (support): Member installed at right angles to a chord or web members of trusses for alignment and support.

Lateral load (force): Side-to-side force acting on a structure.

Layer (plywood): Single veneer ply, or two or more plies, laminated with a parallel grain direction.

Ledger: Horizontal member attached to a wall or girder to support joists or rafters.

Life safety design: Structural design formulated to prevent the collapse of a building due to earthquake or fire.

Lintel (header): Horizontal member installed over an opening in a wall to support the wall construction above.

Load, dead: The weight of all permanent structural and nonstructural components of a building, such as walls, floors, roofs and fixed service equipment.

Load, factored: Load multiplied by appropriate load factors, used to proportion members by strength design method.

Load, live: Any load that isn't permanent, such as people and temporary construction loads.

Load, service: Live and dead loads.

Lumber: Wood that has been sawed, planed, and cross-cut to length.

M

Machine stress-rated lumber (MSR): Mechanically-graded lumber 2 inches or less thick and at least 2 inches wide used in formwork and shoring. Also called *machine-evaluated lumber* (MEL).

MBF: Thousand board feet.

MBM: Thousand (feet) board measure.

Mechanically-laminated: Laminated wood structural member held together with mechanical fasteners.

Modulus of elasticity: Ratio of normal stress to corresponding strain from tensile or compressive stresses below proportional limit of the material.

Modulus of rupture: Maximum bending stress.

Moisture content (wood): Weight of water in wood divided by its dry weight.

Moment of inertia (cross section of a beam): Property of a structural shape's measure of ability to resist changing shape, or to indicate member's strength.

Mud sill: Wood member (usually made of treated wood or redwood) bolted to the top of a foundation to support floor joists. Also called *wood sill or sill plate.*

Multi-tier shoring: Single-post shores used in two or more tiers to increase the height of the shoring platform.

N

NCMA: National Concrete Masonry Association.

NDS: National Design Specifications for Wood.

NER: National Evaluation Report.

Neutral axis: In the cross section of a beam, the location within a board where there is neither tension nor compression stress.

Nominal size (dimension): Rough lumber size before finishing or surfacing.

Nominal span: Horizontal distance between the edges of the supports of a beam or truss.

Nominal thickness (plywood): Full designated thickness of plywood before sanding.

Non-bearing wall: Wall or partition that only carries its own weight.

NRMCA: National Ready Mixed Concrete Association.

O

Occupancy: Purpose for which a building is to be used.

On center (oc): Distance between the centers of adjacent repetitive structural members.

P

Pan: A prefabricated form unit, most commonly of steel, used in concrete floor joist construction.

Parapet wall: Part of an exterior wall above a roofline.

Particleboard: Mat-formed panel made of wood particles or a combination of wood particles and wood fibers bonded together with synthetic resins and used in formwork.

Partition: Interior wall that subdivides building spaces.

PCA: Portland Cement Association.

Pedestal: Upright compression member with a ratio of unsupported height to average least lateral dimension of 3 or less.

Pier: Masonry or concrete column used to support a beam.

Pier block: Preformed concrete footing that supports a post.

Plain concrete: Structural concrete with no reinforcement, or with less reinforcement than minimum amount specified for reinforced concrete.

Plain reinforcement: Reinforcement that doesn't conform to definition of deformed reinforcement. Usually, smooth bars without bumps.

Plumb: True and level on a vertical plane.

Plumb bob: Metal weight suspended from a cord used to establish a vertical line.

Ply: Single veneer lamina in a glued plywood panel. Also, the number of thicknesses of veneer in a plywood panel or laminated member.

Plyform: Plastic-coated plywood used for forming concrete.

Pneumatically-driven fasteners: Air-driven nails, staples, or spikes.

Post (column): Vertical load-bearing structural member.

Post-tensioning: Method of prestressing tendons in concrete after the concrete has hardened.

Precast concrete: Structural concrete element that is cast and then placed in its final position in a structure.

Prestressed concrete: Structural concrete in which internal compressive stresses have been introduced by the steel tendons to reduce potential tensile stresses in concrete resulting from loads.

Pre-tensioning: Method of prestressing tendons in concrete before the concrete is placed.

Purlin: Horizontal member that acts as a beam and supports common rafters or ceiling joists.

R

Racking: Twisting movement that can distort a framework.

Reaction: Load transmitted from a beam or truss to a support.

Registered design professional: Any architect or engineer registered or licensed to design a project in a given state according to the state's professional registration laws.

Reinforced concrete: Structural concrete that is reinforced with no less than the minimum amounts of prestressing tendons or nonprestressed reinforcement specified in the building code.

Reshores (reshoring): Shores placed snugly under a concrete slab or other structural member after the original forms and shores have been removed.

Retrofit: To add additional bracing, anchoring, or any improvement to a completed structure.

S

Scab: Small piece of wood fastened to two formwork members to secure a butt joint.

Scaffolding: Elevated platform erected to support workers, tools, and materials.

Seasoning (wood): Drying lumber by exposure to air and sun or by kiln.

Section modulus: Property of the shape of a structural member that indicates its strength; the moment of inertia divided by the distance from the neutral axis to the extreme fiber of the section.

Select structural lumber: High-quality lumber, free of characteristics that impair strength or stiffness.

Shear load: Side-to-side force(s) acting on a structure.

Shear stress: Stress that tends to keep two adjoining planes of a body from sliding on each other when two equal and opposite parallel forces act on them in opposite direction.

Shear wall: A wall designed to resist horizontal loads.

Sheathing: Structural covering applied to the outside surface of wall or roof frame. Also called *sheeting.*

Shim: Long narrow piece, usually of wood, to adjust spacing.

Shop drawing: Drawing, diagram, illustration, schedule, and other data prepared by a contractor, manufacturer, fabricator, supplier, or distributor to show some portion of the work.

Shore: Temporary vertical or inclined member that supports formwork and fresh concrete until the structure has developed full strength.

Siding: Finish covering applied to the outer side of the exterior walls of a frame building.

Sill plate: Lowest horizontal member of a wall frame bolted to the foundation. This member is also called a *sole plate*, or *mud sill.*

Sleepers: Treated wood nailers attached to a concrete slab, providing a nailing base for flooring.

Snow load: Load on a building resulting from accumulated snow.

Soffit: Underside of a structural member of a building, such as a beam.

Soldier: Vertical wales used to strengthen and align forms.

Span: Horizontal distance between supports.

Span rating (plywood): Pair of numbers stamped on plywood sheathing to indicate its span capabilities over roofs and floors. Also called an *identification index.*

Specifications: Technical descriptions of a project regarding materials, equipment, construction systems, standards, and workmanship.

Spiral reinforcement: Continuously wound reinforcement in the form of a cylindrical helix.

Split: Separation of wood fibers caused by external forces.

Stirrup: (1) A metal bar used in concrete reinforcement; (2) A metal supporting strap used to hold one end of a beam or joist and connect it to another structural member.

Strength, design: Nominal strength multiplied by a strength-reduction factor f.

Stress: Intensity of force per unit area.

Strike: To lower or remove formwork or centering.

Stringer: A beam that supports floor or deck sheathing.

Stripping: Disassembling forming and shoring, usually for reuse.

Strips: Boards less than 6 inches wide.

Structural concrete: All concrete used for structural purposes, including plain and reinforced concrete.

Strut: Horizontal or inclined compression member.

Stud: Vertical wood or metal member that supports the sheathing in wall forming.

Subcontractor: Individual, firm, or corporation having a direct contract with a general contractor or subcontractor for the performance of part of a project.

Sway brace: Diagonal brace used to resist wind or other lateral forces. Also see *X-brace*.

T

T&G: Tongue-and-groove.

T-head: Top of a shore with a braced horizontal member projecting on two sides, forming a T-shaped assembly.

Teco nail: Nail used to install hangers.

Tell-tale: Any device designed to indicate the movement of formwork.

Template: Thin plate or board frame used as a guide in positioning or spacing form parts, reinforcement, anchors, etc.

Tendon: Steel element such as wire, cable, bar, rod, or strand, or bundle of such elements, used to prestress concrete.

Tensile strength: Resistance to tensile forces.

Tensile stress: Pulling force over a unit area expressed in psi, or tensile force divided by the tensile area.

Tension: Force exerted on a structural member that tends to pull it apart or elongate it.

Tie: Loop of reinforcing bar or wire enclosing lengthwise reinforcement. A continuous wound bar or wire in the form of a circle, rectangle, or other polygon shape.

Tie wire: Metal wires used to hold opposing forms in position.

Timbers: Lumber 5 or more inches in the least dimension, including beams, stringers, posts, sills, girders, and purlins.

Tributary area (domain): Roof, floor or wall area that contributes to the load on a structural member.

Truss: Shop-fabricated frame used as a roof support.

U

Ultimate stress: The maximum stress that a material can stand before it breaks apart.

Underlayment: Sheathing material used as a base for finish flooring or carpet.

Uniform load: Load that is equally distributed over a given length of a beam, and is usually expressed as pounds per lineal foot (plf).

Unit stress: Amount of stress on 1 square inch of a material.

Uplift: Force(s) acting to lift a structure caused by earthquake or wind.

W

Wale (waler): Horizontal member used to align and brace studs on concrete forms.

Wane: Lumber defect located near the edge or corner of a board caused by the lack of bark.

Wood shear panel: Wood floor, roof, or wall frame sheathed to act as a shear wall or diaphragm.

Work: Construction required under the contract documents.

Working stress: Unit stress that experience has shown to be safe for a material.

WRI: Wire Reinforcing Institute.

X

X-brace: Paired set of tension sway braces made of steel rods, angles, or other members used to resist sideways overturning from lateral forces on a structure. Also see *Sway brace*.

Y

Yard lumber: Wood members used for common construction.

Yield strength: Specified minimum yield strength or yield point of reinforcement, in psi.

Yoke: Tie or clamping device placed around column forms or over the top of wall or footing forms to keep them from spreading due to lateral concrete pressure.

Index

Practical References for Builders

Construction Estimating Reference Data

Provides the 300 most useful manhour tables for practically every item of construction. Labor requirements are listed for sitework, concrete work, masonry, steel, carpentry, thermal and moisture protection, doors and windows, finishes, mechanical and electrical. Each section details the work being estimated and gives appropriate crew size and equipment needed. Includes a CD-ROM with an electronic version of the book with *National Estimator*, a stand-alone *Windows*™ estimating program, plus an interactive multimedia video that shows how to use the disk to compile construction cost estimates. **432 pages, 11 x 8¹/₂, $39.50**

Concrete Construction & Estimating

Explains how to estimate the quantity of labor and materials needed, plan the job, erect fiberglass, steel, or prefabricated forms, install shores and scaffolding, handle the concrete into place, set joints, finish and cure the concrete. Full of practical reference data, cost estimates, and examples. **571 pages, 5¹/₂ x 8¹/₂, $25.00**

Commercial Electrical Wiring

Make the transition from residential to commercial electrical work. Here are wiring methods, spec reading tips, load calculations and everything you need for making the transition to commercial work: commercial construction documents, load calculations, electric services, transformers, overcurrent protection, wiring methods, raceway, boxes and fittings, wiring devices, conductors, electric motors, relays and motor controllers, special occupancies, and safety requirements. This book is written to help any electrician break into the lucrative field of commercial electrical work. Updated to the 1999 *NEC*. **320 pages, 8¹/₂ x 11, $36.50**

ADA Accessibility Guidelines for Buildings and Facilities With ADA Technical Assistance Manuals

All buildings, both old and new, must comply with the Americans with Disabilities Act. This book provides a practical, step-by-step checklist to assure compliance for all types of buildings, from offices and retail to hospitals and public facilities. Here you'll find diagrams, forms and 150 pages of checklists to make sure that what you build or remodel will pass ADA guidelines. Also includes the Technical Assistance Manuals for the ADA Titles II and III. **350 pages, 8¹/₂ x 11, $69.95**

Estimating with Microsoft *Excel*

Most builders estimate with *Excel* because it's easy to learn, quick to use, and can be customized to your style of estimating. Here you'll find step-by-step how to create your own customized automated spreadsheet estimating program for use with *Excel*. You'll learn how to use the magic of *Excel* in creating detail sheets, cost breakdown summaries, and linking. You can even create your own macros. Includes a CD-ROM that illustrates examples in the book and provides you with templates you can use to set up your own estimating system. **148 pages, 8¹/₂ x 11, $49.95**

Basic Construction Management: The Superintendent's Job, 4th Edition

Today's construction projects are more complex than ever. Managing these projects has also become more complex. This perennial NAHB bestseller, now in its fourth edition, addresses the issues facing today's construction manager. New managers can use this as a great training tool. Experienced superintendents can brush up on the latest techniques and technologies. **198 pages, 8¹/₂ x 11, $40.00**

Building Layout

Shows how to use a transit to locate a building correctly on the lot, plan proper grades with minimum excavation, find utility lines and easements, establish correct elevations, lay out accurate foundations, and set correct floor heights. Explains how to plan sewer connections, level a foundation that's out of level, use a story pole and batterboards, work on steep sites, and minimize excavation costs. **240 pages, 5¹/₂ x 8¹/₂, $19.00**

Basic Lumber Engineering for Builders

Beam and lumber requirements for many jobs aren't always clear, especially with changing building codes and lumber products. Most of the time you rely on your own "rules of thumb" when figuring spans or lumber engineering. This book can help you fill the gap between what you can find in the building code span tables and what you need to pay a certified engineer to do. With its large, clear illustrations and examples, this book shows you how to figure stresses for pre-engineered wood or wood structural members, how to calculate loads, and how to design your own girders, joists and beams. Included FREE with the book — an easy-to-use limited version of NorthBridge Software's *Wood Beam Sizing* program. **272 pages, 8¹/₂ x 11, $38.00**

Blueprint Reading for the Building Trades

How to read and understand construction documents, blueprints, and schedules. Includes layouts of structural, mechanical, HVAC and electrical drawings. Shows how to interpret sectional views, follow diagrams and schematics, and covers common problems with construction specifications. **192 pages, 5¹/₂ x 8¹/₂, $14.75**

BNI Public Works Costbook, 2001

This is the only book of its kind for public works construction. Here you'll find labor and material prices for most public works and infrastructure projects: roads and streets, utilities, street lighting, manholes, and much more. Includes manhour data and a 200-city geographic modifier chart. Includes FREE estimating software and data. **450 pages, 8¹/₂ x 11, $79.95**

Build Smarter with Alternative Materials

New building products are coming out almost every week. Some of them may become new standards, as sheetrock replaced lath and plaster some years ago. Others are little more than a gimmick. To write this manual, the author researched hundreds of products that have come on the market in recent years. The ones he describes in this book will do the job better, creating a superior, longer-lasting finished product, and in many cases also save you time and money. Some are made with recycled products — a good selling point with many customers. But most of all, they give you choices, so you can give your customers choices. In this book, you'll find materials for almost all areas of constructing a house, from the ground up. For each product described, you'll learn where you can get it, where to use it, what benefits it provides, any disadvantages, and how to install it — including tips from the author. And to help you price your jobs, each description ends with manhours — for both the first time you install it, and after you've done it a few times. **336 pages, 8¹/₂ x 11, $34.75**

Builder's Guide to Accounting Revised

Step-by-step, easy-to-follow guidelines for setting up and maintaining records for your building business. This practical, newly-revised guide to all accounting methods shows how to meet state and federal accounting requirements, explains the new depreciation rules, and describes how the Tax Reform Act can affect the way you keep records. Full of charts, diagrams, simple directions and examples, to help you keep track of where your money is going. Recommended reading for many state contractor's exams. **320 pages, 8¹/₂ x 11, $26.50**

CD Estimator

If your computer has *Windows*™ and a CD-ROM drive, *CD Estimator* puts at your fingertips 85,000 construction costs for new construction, remodeling, renovation & insurance repair, electrical, plumbing, HVAC and painting. You'll also have the *National Estimator* program — a stand-alone estimating program for *Windows*™ that *Remodeling* magazine called a "computer wiz." Quarterly cost updates are available at no charge on the Internet. To help you create professional-looking estimates, the disk includes over 40 construction estimating and bidding forms in a format that's perfect for nearly any word processing or spreadsheet program for Windows. And to top it off, a 70-minute interactive video teaches you how to use this CD-ROM to estimate construction costs. **CD Estimator is $68.50**

Steel-Frame House Construction

Framing with steel has obvious advantages over wood, yet building with steel requires new skills that can present challenges to the wood builder. This new book explains the secrets of steel framing techniques for building homes, whether pre-engineered or built stick by stick. It shows you the techniques, the tools, the materials, and how you can make it happen. Includes hundreds of photos and illustrations, plus a CD-ROM with steel framing details.
320 pages, 8¹/₂ x 11, $39.75

Electrician's Exam Preparation Guide

Need help in passing the apprentice, journeyman, or master electrician's exam? This is a book of questions and answers based on actual electrician's exams over the last few years. Almost a thousand multiple-choice questions — exactly the type you'll find on the exam — cover every area of electrical installation: electrical drawings, services and systems, transformers, capacitors, distribution equipment, branch circuits, feeders, calculations, measuring and testing, and more. It gives you the correct answer, an explanation, and where to find it in the latest *NEC*. Also tells how to apply for the test, how best to study, and what to expect on examination day.
352 pages, 8¹/₂ x 11, $32.00

Basic Engineering for Builders

If you've ever been stumped by an engineering problem on the job, yet wanted to avoid the expense of hiring a qualified engineer, you should have this book. Here you'll find engineering principles explained in non-technical language and practical methods for applying them on the job. With the help of this book you'll be able to understand engineering functions in the plans and how to meet the requirements, how to get permits issued without the help of an engineer, and anticipate requirements for concrete, steel, wood and masonry. See why you sometimes have to hire an engineer and what you can undertake yourself: surveying, concrete, lumber loads and stresses, steel, masonry, plumbing, and HVAC systems. This book is designed to help the builder save money by understanding engineering principles that you can incorporate into the jobs you bid.
400 pages, 8¹/₂ x 11, $34.00

Estimating Home Building Costs

Estimate every phase of residential construction from site costs to the profit margin you include in your bid. Shows how to keep track of man-hours and make accurate labor cost estimates for footings, foundations, framing and sheathing finishes, electrical, plumbing, and more. Provides and explains sample cost estimate worksheets with complete instructions for each job phase. **320 pages, 5¹/₂ x 8¹/₂, $17.00**

Estimating Excavation

How to calculate the amount of dirt you'll have to move and the cost of owning and operating the machines you'll do it with. Detailed, step-by-step instructions on how to assign bid prices to each part of the job, including labor and equipment costs. Also, the best ways to set up an organized and logical estimating system, take off from contour maps, estimate quantities in irregular areas, and figure your overhead.
448 pages, 8¹/₂ x 11, $39.50

Handbook of Construction Contracting

Volume 1: Everything you need to know to start and run your construction business; the pros and cons of each type of contracting, the records you'll need to keep, and how to read and understand house plans and specs so you find any problems before the actual work begins. All aspects of construction are covered in detail, including all-weather wood foundations, practical math for the job site, and elementary surveying.
416 pages, 8¹/₂ x 11, $32.75

Volume 2: Everything you need to know to keep your construction business profitable; different methods of estimating, keeping and controlling costs, estimating excavation, concrete, masonry, rough carpentry, roof covering, insulation, doors and windows, exterior finishes, specialty finishes, scheduling work flow, managing workers, advertising and sales, spec building and land development, and selecting the best legal structure for your business. **320 pages, 8¹/₂ x 11, $30.75**

National Building Cost Manual

Square foot costs for residential, commercial, industrial, and farm buildings. Quickly work up a reliable budget estimate based on actual materials and design features, area, shape, wall height, number of floors, and support requirements. Includes all the important variables that can make any building unique from a cost standpoint.
240 pages, 8¹/₂ x 11, $23.00. Revised annually

Home Inspection Handbook

Every area you need to check in a home inspection — especially in older homes. Twenty complete inspection checklists: building site, foundation and basement, structural, bathrooms, chimneys and flues, ceilings, interior & exterior finishes, electrical, plumbing, HVAC, insects, vermin and decay, and more. Also includes information on starting and running your own home inspection business. **324 pages, 5¹/₂ x 8¹/₂, $24.95**

How to Succeed With Your Own Construction Business

Everything you need to start your own construction business: setting up the paperwork, finding the work, advertising, using contracts, dealing with lenders, estimating, scheduling, finding and keeping good employees, keeping the books, and coping with success. If you're considering starting your own construction business, all the knowledge, tips, and blank forms you need are here. **336 pages, 8¹/₂ x 11, $28.50**

Illustrated Guide to the 1999 *National Electrical Code*

This fully-illustrated guide offers a quick and easy visual reference for installing electrical systems. Whether you're installing a new system or repairing an old one, you'll appreciate the simple explanations written by a code expert, and the detailed, intricately-drawn and labeled diagrams. A real time-saver when it comes to deciphering the current *NEC*.
360 pages, 8¹/₂ x 11, $38.75

Estimating Electrical Construction

Like taking a class in how to estimate materials and labor for residential and commercial electrical construction. Written by an A.S.P.E. National Estimator of the Year, it teaches you how to use labor units, the plan take-off, and the bid summary to make an accurate estimate, how to deal with suppliers, use pricing sheets, and modify labor units. Provides extensive labor unit tables and blank forms for your next electrical job.
272 pages, 8¹/₂ x 11, $19.00

Masonry & Concrete Construction Revised

This is the revised edition of the popular manual, with updated information on everything from on-site pre-planning and layout through the construction of footings, foundations, walls, fireplaces and chimneys. There's an added appendix on safety regulations, with all the applicable OSHA sections pulled together into one handy condensed reference. There's new information on concrete, masonry and seismic reinforcement. Plus improved estimating techniques to help you win more construction bids. The emphasis is on integrating new techniques and improved materials with the tried-and-true methods. Includes information on cement and mortar types, mixes, coloring agents and additives, and suggestions on when, where and how to use them; calculating footing and foundation loads, with tables and formulas to use as references; forming materials and forming systems; pouring and reinforcing concrete slabs and flatwork; block and brick wall construction, including seismic requirements; crack control, masonry veneer construction, brick floors and pavements, including design considerations and materials; and cleaning, painting and repairing all types of masonry.
304 pages, 8¹/₂ x 11, $28.50

Excavation & Grading Handbook Revised

Explains how to handle all excavation, grading, compaction, paving and pipeline work: setting cut and fill stakes (with bubble and laser levels), working in rock, unsuitable material or mud, passing compaction tests, trenching around utility lines, setting grade pins and string line, removing or laying asphaltic concrete, widening roads, cutting channels, installing water, sewer, and drainage pipe. This is the completely revised edition of the popular guide used by over 25,000 excavation contractors.
384 pages, 5¹/₂ x 8¹/₂, $22.75

Markup & Profit: A Contractor's Guide

In order to succeed in a construction business, you have to be able to price your jobs to cover all labor, material and overhead expenses, and make a decent profit. The problem is knowing what markup to use. You don't want to lose jobs because you charge too much, and you don't want to work for free because you've charged too little. If you know how to calculate markup, you can apply it to your job costs to find the right sales price for your work. This book gives you tried and tested formulas, with step-by-step instructions and easy-to-follow examples, so you can easily figure the markup that's right for your business. Includes a CD-ROM with forms and checklists for your use. **320 pages, 8¹/₂ x 11, $32.50**

Concrete & Formwork

This practical manual has all the information you need to select and pour the right mix for the job, lay out the structure, choose the right form materials, design and build the forms, and finish and cure the concrete, Nearly 100 pages of step-by-step instructions show how to construct and erect most types of site-fabricated wood forms used in residential construction. **176 pages, 8¹/₂ x 11, $17.75**

National Construction Estimator

Current building costs for residential, commercial, and industrial construction. Estimated prices for every common building material. Provides man-hours, recommended crew, and gives the labor cost for installation. Includes a CD-ROM with an electronic version of the book with *National Estimator*, a stand-alone *Windows*™ estimating program, plus an interactive multimedia video that shows how to use the disk to compile construction cost estimates. **616 pages, 8¹/₂ x 11, $47.50. Revised annually**

Construction Forms & Contracts

125 forms you can copy and use — or load into your computer (from the FREE disk enclosed). Then you can customize the forms to fit your company, fill them out, and print. Loads into *Word* for *Windows*™, *Lotus 1-2-3*, *WordPerfect*, *Works*, or *Excel* programs. You'll find forms covering accounting, estimating, fieldwork, contracts, and general office. Each form comes with complete instructions on when to use it and how to fill it out. These forms were designed, tested and used by contractors, and will help keep your business organized, profitable and out of legal, accounting and collection troubles. Includes a CD-ROM for *Windows*™ and Mac. **400 pages, 8¹/₂ x 11, $41.75**

Contracting in All 50 States

Every state has its own licensing requirements that you must meet to do business there. These are usually written exams, financial requirements, and letters of reference. This book shows how to get a building, mechanical or specialty contractor's license, qualify for DOT work, and register as an out-of-state corporation, for every state in the U.S. It lists addresses, phone numbers, application fees, requirements, where an exam is required, what's covered on the exam and how much weight each area of construction is given on the exam. You'll find just about everything you need to know in order to apply for your out-of-state license. **416 pages, 8¹/₂ x 11, $36.00**

Contractor's Growth & Profit Guide

Step-by-step instructions for planning growth and prosperity in a construction contracting or subcontracting company. Explains how to prepare a business plan: select reasonable goals, draft a market expansion plan, make income forecasts and expense budgets, and project cash flow. You'll learn everything that most lenders and investors require, as well as the best way to organize your business. **336 pages, 5¹/₂ x 8¹/₂, $19.00**

The Contractor's Legal Kit

Stop "eating" the costs of bad designs, hidden conditions, and job surprises. Set ground rules that assign those costs to the rightful party ahead of time. And it's all in plain English, not "legalese." For less than the cost of an hour with a lawyer you'll learn the exclusions to put in your agreements, why your insurance company may pay for your legal defense, how to avoid liability for injuries to your sub and his employees or damages they cause, how to collect on lawsuits you win, and much more. It also includes a FREE computer disk with contracts and forms you can customize for your own use. **352 pages, 8¹/₂ x 11, $59.95**

Contractor's Guide to QuickBooks Pro 2000

This user-friendly manual walks you through QuickBooks Pro's detailed setup procedure and explains step-by-step how to create a first-rate accounting system. You'll learn in days, rather than weeks, how to use QuickBooks Pro to get your contracting business organized, with simple, fast accounting procedures. On the CD included with the book you'll find a QuickBooks Pro file preconfigured for a construction company (you drag it over onto your computer and plug in your own company's data). You'll also get a complete estimating program, including a database, and a job costing program that lets you export your estimates to QuickBooks Pro. It even includes many useful construction forms to use in your business. **304 pages, 8¹/₂ x 11, $44.50**

Contractor's Guide to the Building Code Revised

This new edition was written in collaboration with the International Conference of Building Officials, writers of the code. It explains in plain English exactly what the latest edition of the *Uniform Building Code* requires. Based on the 1997 code, it explains the changes and what they mean for the builder. Also covers the *Uniform Mechanical Code* and the *Uniform Plumbing Code*. Shows how to design and construct residential and light commercial buildings that'll pass inspection the first time. Suggests how to work with an inspector to minimize construction costs, what common building shortcuts are likely to be cited, and where exceptions may be granted. **320 pages, 8¹/₂ x 11, $39.00**

Craftsman's Illustrated Dictionary of Construction Terms

Almost everything you could possibly want to know about any word or technique in construction. Hundreds of up-to-date construction terms, materials, drawings and pictures with detailed, illustrated articles describing equipment and methods. Terms and techniques are explained or illustrated in vivid detail. Use this valuable reference to check spelling, find clear, concise definitions of construction terms used on plans and construction documents, or learn about 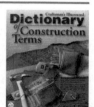 little-known tools, equipment, tests and methods used in the building industry. It's all here. **416 pages, 8¹/₂ x 11, $36.00**

Contractor's Index to the 1997 *Uniform Building Code*

Finally, there's a common-sense index that helps you quickly and easily find the section you're looking for in the *UBC*. It lists topics under the names builders actually use in construction. Best of all, it gives the full section number and the actual page in the *UBC* where you'll find it. If you need to know the requirements for windows in exit access corridor walls, just look under *Windows*™. You'll find the requirements you need are in Section 1004.3.4.3.2.2 in the *UBC* — on page 115. This practical index was written by a former builder and building inspector who knows the *UBC* from both perspectives. If you hate to spend valuable time hunting through pages of fine print for the information you need, this is the book for you. **192 pages, 8¹/₂ x 11, paperback edition, $26.00**
192 pages, 8¹/₂ x 11, looseleaf edition, $29.00.

Contractor's Survival Manual

How to survive hard times and succeed during the up cycles. Shows what to do when the bills can't be paid, finding money and buying time, transferring debt, and all the alternatives to bankruptcy. Explains how to build profits, avoid problems in zoning and permits, taxes, time-keeping, and payroll. Unconventional advice on how to invest in inflation, get high appraisals, trade and postpone income, and stay hip-deep in profitable work. **160 pages, 8¹/₂ x 11, $22.25**

Troubleshooting Guide to Residential Construction

How to solve practically every construction problem — before it happens to you! With this book you'll learn from the mistakes other builders made as they faced 63 typical residential construction problems. Filled with clear photos and drawings that explain how to enhance your reputation as well as your bottom line by avoiding problems that plague most builders. Shows how to avoid, or fix, problems ranging from defective slabs, walls and ceilings, through roofing, plumbing & HVAC, to paint. **304 pages, 8¹/₂ x 11, $32.50**

Getting Financing & Developing Land

Developing land is a major leap for most builders — yet that's where the big money is made. This book gives you the practical knowledge you need to make that leap. Learn how to prepare a market study, select a building site, obtain financing, guide your plans through approval, then control your building costs so you can ensure yourself a good profit. Includes a CD-ROM with forms, checklists, and a sample business plan you can customize and use to help you sell your idea to lenders and investors. **232 pages, 8¹/₂ x 11, $39.00**

Contractor's Year-Round Tax Guide Revised

How to set up and run your construction business to minimize taxes: corporate tax strategy and how to use it to your advantage, and what you should be aware of in contracts with others. Covers tax shelters for builders, write-offs and investments that will reduce your taxes, accounting methods that are best for contractors, and what the I.R.S. allows and what it often questions. **192 pages, 8¹/₂ x 11, $26.50**

Cost Records for Construction Estimating

How to organize and use cost information from jobs just completed to make more accurate estimates in the future. Explains how to keep the records you need to track costs for sitework, footings, foundations, framing, interior finish, siding and trim, masonry, and subcontract expense. Provides sample forms. **208 pages, 8¹/₂ x 11, $15.75**

Residential Electrical Estimating

A fast, accurate pricing system proven on over 1000 residential jobs. Using the manhours provided, combined with material prices from your wholesaler, you quickly work up estimates based on degree of difficulty. These manhours come from a working electrical contractor's records — not some pricing agency. You'll find prices for every type of electrical job you're likely to estimate — from service entrances to ceiling fans. **320 pages, 8¹/₂ x 11, $29.00**

National Renovation & Insurance Repair Estimator

Current prices in dollars and cents for hard-to-find items needed on most insurance, repair, remodeling, and renovation jobs. All price items include labor, material, and equipment breakouts, plus special charts that tell you exactly how these costs are calculated. Includes a CD-ROM with an electronic version of the book with *National Estimator*, a stand-alone *Windows*™ estimating program, plus an interactive multimedia video that shows how to use the disk to compile construction cost estimates. **568 pages, 8¹/₂ x 11, $49.50. Revised annually**

National Repair & Remodeling Estimator

The complete pricing guide for dwelling reconstruction costs. Reliable, specific data you can apply on every repair and remodeling job. Up-to-date material costs and labor figures based on thousands of jobs across the country. Provides recommended crew sizes; average production rates; exact material, equipment, and labor costs; a total unit cost and a total price including overhead and profit. Separate listings for high- and low-volume builders, so prices shown are specific for any size business. Estimating tips specific to repair and remodeling work to make your bids complete, realistic, and profitable. Includes a CD-ROM with an electronic version of the book with *National Estimator*, a stand-alone *Windows*™ estimating program, plus an interactive multimedia video that shows how to use the disk to compile construction cost estimates. **296 pages, 8¹/₂ x 11, $48.50. Revised annually**

Plumber's Handbook Revised

This new edition shows what will and won't pass inspection in drainage, vent, and waste piping, septic tanks, water supply, graywater recycling systems, pools and spas, fire protection, and gas piping systems. All tables, standards, and specifications are completely up-to-date with recent plumbing code changes. Covers common layouts for residential work, how to size piping, select and hang fixtures, practical recommendations, and trade tips. It's the approved reference for the plumbing contractor's exam in many states. Includes an extensive set of multiple choice questions after each chapter, and in the back of the book, the answers and explanations. Also in the back of the book, a full sample plumber's exam. **352 pages, 8¹/₂ x 11, $32.00**

Profits in Buying & Renovating Homes

Step-by-step instructions for selecting, repairing, improving, and selling highly profitable "fixer-uppers." Shows which price ranges offer the highest profit-to-investment ratios, which neighborhoods offer the best return, practical directions for repairs, and tips on dealing with buyers, sellers, and real estate agents. Shows you how to determine your profit before you buy, what "bargains" to avoid, and how to make simple, profitable, inexpensive upgrades. **304 pages, 8¹/₂ x 11, $19.75**

Professional Kitchen Design

Remodeling kitchens requires a "special" touch — one that blends artistic flair with function to create a kitchen with charm and personality as well as one that is easy to work in. Here you'll find how to make the best use of the space available in any kitchen design job, as well as tips and lessons on how to design one-wall, two-wall, L-shaped, U-shaped, peninsula and island kitchens. Also includes what you need to know to run a profitable kitchen design business. **176 pages, 8¹/₂ x 11, $24.50**

Residential Wiring to the 1999 *NEC*

Shows how to install rough and finish wiring in new construction, alterations, and additions. Complete instructions on troubleshooting and repairs. Every subject is referenced to the most recent *National Electrical Code*, and there's 22 pages of the most-needed *NEC* tables to help make your wiring pass inspection — the first time. **352 pages, 5¹/₂ x 8¹/₂, $27.00**

Roof Framing

Shows how to frame any type of roof in common use today, even if you've never framed a roof before. Includes using a pocket calculator to figure any common, hip, valley, or jack rafter length in seconds. Over 400 illustrations cover every measurement and every cut on each type of roof: gable, hip, Dutch, Tudor, gambrel, shed, gazebo, and more. **480 pages, 5¹/₂ x 8¹/₂, $22.00**

Roof Framer's Bible

68 different pitch combinations of "bastard" hip roofs at your fingertips. Don't curse the architect — let this book make you an accomplished master of irregular pitched roof systems. You'll be the envy of your crew, and irregular or "bastard" roofs will be under your command. This rare pocket-sized book comes hardbound with a cloth marker like a true bible. **216 pages, 3³/₄ x 7¹/₂, $24.00**

Simplified Guide to Construction Law

Here you'll find easy-to-read, paragraphed-sized samples of how the courts have viewed common areas of disagreement — and litigation — in the building industry. You'll read about legal problems that real builders have faced, and how the court ruled. This book will tell you what you need to know about contracts, torts, fraud, misrepresentation, warranty and strict liability, construction defects, indemnity, insurance, mechanics liens, bonds and bonding, statutes of limitation, arbitration, and more. These are simplified examples that illustrate not necessarily who is right and who is wrong — but who the law has sided with. **298 pages, 5¹/₂ x 8¹/₂, $29.95**

Builder's Guide to Room Additions

How to tackle problems that are unique to additions, such as requirements for basement conversions, reinforcing ceiling joists for second-story conversions, handling problems in attic conversions, what's required for footings, foundations, and slabs, how to design the best bathroom for the space, and much more. Besides actual construction, you'll even find help in designing, planning, and estimating your room addition jobs. **352 pages, 8¹/₂ x 11, $27.25**

Roofing Construction & Estimating

Installation, repair and estimating for nearly every type of roof covering available today in residential and commercial structures: asphalt shingles, roll roofing, wood shingles and shakes, clay tile, slate, metal, built-up, and elastomeric. Covers sheathing and underlayment techniques, as well as secrets for installing leakproof valleys. Many estimating tips help you minimize waste, as well as insure a profit on every job. Troubleshooting techniques help you identify the true source of most leaks. Over 300 large, clear illustrations help you find the answer to just about all your roofing questions. **432 pages, 8¹/₂ x 11, $35.00**

Building Contractor's Exam Preparation Guide

Passing today's contractor's exams can be a major task. This book shows you how to study, how questions are likely to be worded, and the kinds of choices usually given for answers. Includes sample questions from actual state, county, and city examinations, plus a sample exam to practice on. This book isn't a substitute for the study material that your testing board recommends, but it will help prepare you for the types of questions — and their correct answers — that are likely to appear on the actual exam. Knowing how to answer these questions, as well as what to expect from the exam, can greatly increase your chances of passing. **320 pages, 8¹/₂ x 11, $35.00**

Rough Framing Carpentry

If you'd like to make good money working outdoors as a framer, this is the book for you. Here you'll find shortcuts to laying out studs; speed cutting blocks, trimmers and plates by eye; quickly building and blocking rake walls; installing ceiling backing, ceiling joists, and truss joists; cutting and assembling hip trusses and California fills; arches and drop ceilings — all with production line procedures that save you time and help you make more money. Over 100 on-the-job photos of how to do it right and what can go wrong. **304 pages, 8¹/₂ x 11, $26.50**

Commercial Metal Stud Framing

Framing commercial jobs can be more lucrative than residential work. But most commercial jobs require some form of metal stud framing. This book teaches step-by-step, with hundreds of job site photos, high-speed metal stud framing in commercial construction. It describes the special tools you'll need and how to use them effectively, and the material and equipment you'll be working with. You'll find the shortcuts, tips and tricks-of-the-trade that take most steel frames years on the job to discover. Shows how to set up a crew to maintain a rhythm that will speed progress faster than any wood framing job. If you've framed with wood, this book will teach you how to be one of the few top-notch metal stud framers. **208 pages, 8¹/₂ x 11, $45.00**

Residential Concrete

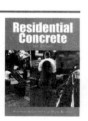

Filled with illustrations , diagrams, and photographs, this essential resource covers building layout, forming for curbs, slabs, and walls. It explains ordering, placing and finishing concrete at the jobsite. It explains the complexities of working with admixtures, jointing, reinforcing and curing. Also shows how to use decorative colors, textures, and finishes. Shows how to prevent problems in concrete and how to repair them if they happen. **112 pages, 8¹/₂ x 11, $32.00.**

Craftsman Book Company
6058 Corte del Cedro
P.O. Box 6500
Carlsbad, CA 92018

☎ 24 hour order line
1-800-829-8123
Fax (760) 438-0398

In A Hurry?
We accept phone orders charged to your

○ Visa, ○ MasterCard, ○ Discover or ○ American Express

Card#_____

Exp.date_____Initials_____

Tax Deductible: Treasury regulations make these references tax deductible when used in your work. Save the canceled check or charge card statement as your receipt.

Name_____

Company_____

Address_____

City/State/Zip_____
○ This is a residence

Total enclosed_____(In California add 7.25% tax

We pay shipping when your check covers your order in full.

Order online http://www.craftsman-book.com
Free on the Internet! Download any of Craftsman's estimating costbooks for a 30-day free trial! http://costbook.com

10-Day Money Back Guarantee

○ 69.95 ADA Accessibility Guidelines for Buildings and Facilities With ADA Technical Assistance Manuals
○ 40.00 Basic Construction Management: The Superintendent's Job, 4th Edition
○ 34.00 Basic Engineering for Builders
○ 38.00 Basic Lumber Engineering for Builders
○ 14.75 Blueprint Reading for the Building Trades
○ 79.95 BNI Public Works Costbook, 2001
○ 34.75 Build Smarter with Alternative Materials
○ 26.50 Builder's Guide to Accounting Revised
○ 27.25 Builder's Guide to Room Additions
○ 35.00 Building Contractor's Exam Preparation Guide
○ 19.00 Building Layout
○ 68.50 CD Estimator
○ 36.50 Commercial Electrical Wiring
○ 45.00 Commercial Metal Stud Framing
○ 17.75 Concrete and Formwork
○ 25.00 Concrete Construction & Estimating
○ 39.50 Construction Estimating Reference Data with FREE *National Estimator* on a CD-ROM.
○ 41.75 Construction Forms & Contracts with a CD-ROM for *Windows*™ and Macintosh.
○ 36.00 Contracting in All 50 States
○ 19.00 Contractor's Growth & Profit Guide
○ 44.50 Contractor's Guide to QuickBooks Pro 2000
○ 39.00 Contractor's Guide to the Building Code Revised
○ 26.00 Contractor's Index to the *UBC* — Paperback
○ 29.00 Contractor's Index to the *UBC* — Looseleaf
○ 59.95 Contractor's Legal Kit
○ 22.25 Contractor's Survival Manual
○ 26.50 Contractor's Year-Round Tax Guide Revised
○ 15.75 Cost Records for Construction Estimating
○ 36.00 Craftsman's Illustrated Dictionary of Construction Terms
○ 32.00 Electrician's Exam Preparation Guide
○ 19.00 Estimating Electrical Construction
○ 39.50 Estimating Excavation

○ 17.00 Estimating Home Building Costs
○ 49.95 Estimating with Microsoft *Excel*
○ 22.75 Excavation & Grading Handbook Revised
○ 39.00 Getting Financing & Developing Land
○ 32.75 Handbook of Construction Contracting Volume 1
○ 30.75 Handbook of Construction Contracting Volume 2
○ 24.95 Home Inspection Handbook
○ 28.50 How to Succeed w/Your Own Construction Business
○ 38.75 Illustrated Guide to the 1999 *National Electrical Code*
○ 32.50 Markup & Profit: A Contractor's Guide
○ 28.50 Masonry & Concrete Construction Revised
○ 23.00 National Building Cost Manual
○ 47.50 National Construction Estimator with FREE *National Estimator* on a CD-ROM.
○ 49.50 National Renovation & Insurance Repair Estimator with FREE *National Estimator* on a CD-ROM.
○ 48.50 National Repair & Remodeling Estimator with FREE *National Estimator* on a CD-ROM.
○ 32.00 Plumber's Handbook Revised
○ 24.50 Professional Kitchen Design
○ 19.75 Profits in Buying & Renovating Homes
○ 32.00 Residential Concrete
○ 29.00 Residential Electrical Estimating
○ 27.00 Residential Wiring to the 1999 *NEC*
○ 22.00 Roof Framing
○ 24.00 Roof Framers Bible
○ 35.00 Roofing Construction & Estimating
○ 26.50 Rough Framing Carpentry
○ 29.95 A Simplified Guide to Construction Law
○ 39.75 Steel-Frame House Construction
○ 32.50 Troubleshooting Guide to Residential Construction
○ 39.50 Basic Concrete Engineering for Builders
○ FREE Full Color Catalog

Prices subject to change without notice